Bioarchaeology of Ethnogenesis in the Colonial Southeast

Florida Museum of Natural History: Ripley P. Bullen Series

UNIVERSITY PRESS OF FLORIDA
Florida A&M University, Tallahassee
Florida Atlantic University, Boca Raton
Florida Gulf Coast University, Ft. Myers
Florida International University, Miami
Florida State University, Tallahassee
New College of Florida, Sarasota
University of Central Florida, Orlando
University of Florida, Gainesville
University of North Florida, Jacksonville
University of South Florida, Tampa
University of West Florida, Pensacola

Bioarchaeology of Ethnogenesis in the Colonial Southeast

Christopher M. Stojanowski

University Press of Florida
Gainesville/Tallahassee/Tampa/Boca Raton
Pensacola/Orlando/Miami/Jacksonville/Ft. Myers/Sarasota

Copyright 2010 by Christopher M. Stojanowski
All rights reserved
Printed in the United States of America on acid-free paper

First cloth printing, 2010
First paperback printing, 2013

Library of Congress Cataloging-in-Publication Data
Stojanowski, Christopher M. (Christopher Michael), 1973–
Bioarchaeology of ethnogenesis in the colonial Southeast /
Christopher M. Stojanowski.
p. cm.—(Florida Museum of Natural History : Ripley P. Bullen series)
Includes bibliographical references and index.
ISBN 978-0-8130-3464-5 (cloth: alk. paper)
ISBN 978-0-8130-4903-8 (pbk.)
 1. Indians of North America—Southern States—Antiquities.
2. Indians of North America—Southern States—Ethnic identity.
3. Indians of North America—Southern States—Migrations—History.
4. Ethnicity—Southern States—History. 5. Human biology—Southern States—History. 6. Human population genetics—Southern States—History. 7. Southern States—History—Colonial period, ca. 1600-1775. 8. Southern States—Ethnic relations. 9. Spaniards—Southern States—History. 10. Spain—Colonies—America—History. I. Title.
E78.S65S74 2010
975.'01—dc22   2009047421

The University Press of Florida is the scholarly publishing agency for the State University System of Florida, comprising Florida A&M University, Florida Atlantic University, Florida Gulf Coast University, Florida International University, Florida State University, New College of Florida, University of Central Florida, University of Florida, University of North Florida, University of South Florida, and University of West Florida.

University Press of Florida
15 Northwest 15th Street
Gainesville, FL 32611-2079
http://www.upf.com

For E.

Contents

List of Illustrations   ix

List of Tables   xi

Preface   xiii

Introduction: The Bioarchaeology of Ethnogenesis   1

One   Genetic Landscapes of Spanish Colonial Florida   12

Two   Interpretive Frameworks   39

Three   Ethnogenesis, Social Identity, and Human Biology: A Bridging Model   47

Four   From Tribe to Ethnic Group, or from Ethnic Group to Tribe: Ethnic Demobilization during the Early Seventeenth Century   60

Five   The Liminal Phase of Ethnogenesis: Objectification within the Eastern Woodlands Tribal Zone   78

Six   The Liminal Phase of Ethnogenesis: Practice and the Lived Experience   100

Seven Bridging Histories: Seminole Ethnogenesis Reconsidered   128

Eight Back to the Past: Precontact Biological Integration as Prelude to Colonial Ethnogenesis   154

Nine Parting Comments: The Bioarchaeology of Ethnogenesis   173

Notes   183

References Cited   197

Index   233

Illustrations

Figure 0.1.	Crushing of a cranium due to taphonomic processes	8
Figure 1.1.	Map of La Florida mission locations circa 1650	14
Figure 1.2.	Cultural groups in La Florida	15
Figure 1.3.	Interpreting genetic distance ordinations	21
Figure 1.4.	Interpreting genetic distance ordinations using hypothesized model of microevolutionary change	21
Figure 1.5.	Map of Florida and Georgia with mission period sample locations noted	25
Figure 1.6.	Eigenvector plot for precontact samples	30
Figure 1.7.	Eigenvector plot for early mission period samples	31
Figure 1.8.	Eigenvector plot for late mission period samples	33
Figure 1.9.	Eigenvector plot for early and late mission period samples	34
Figure 2.1.	Ethnogenetic models derived from the colonial ethnographic literature	44
Figure 5.1.	Process of population aggregation during the 17th century	83
Figure 6.1.	Three mission period church cemetery areas	116
Figure 6.2.	The Irene Mound mortuary complex	119

Figure 6.3.	"Burial Ceremonies for a Chief or Priest"	120
Figure 6.4.	Example of a hypoplastic defect	124
Figure 7.1.	Areas of earliest proto-Seminole settlements in Florida and southern Georgia	139
Figure 7.2.	Ethnic mergers and realignments during the 18th century	141
Figure 7.3.	Change in social identity through time in different regions of Florida and Georgia	151
Figure 7.4.	Patterns of biological integration and migration in different regions of Florida and Georgia	152
Figure 8.1.	Locations of samples used in precontact and proto historic regional comparative analysis	158
Figure 8.2.	Principal components plot for early precontact period samples	163
Figure 8.3.	Patterns of biological integration during the early precontact period	164
Figure 8.4.	Principal components plot for late precontact period samples	166
Figure 8.5.	Patterns of biological integration during the late precontact period	167
Figure 8.6.	Principal components plot for protohistoric period samples	168
Figure 8.7.	Patterns of biological integration during the protohistoric period	169

Tables

Table 1.1.	Important Dates in La Florida History 17
Table 1.2.	Summary Information on Mission Period Samples in La Florida 24
Table 1.3.	Late Precontact Skeletal Samples from La Florida 28
Table 1.4.	RMET Output for Precontact Period Samples 30
Table 1.5.	RMET Output for Early Mission Period Samples 32
Table 1.6.	RMET Output for Late Mission Period Samples 33
Table 1.7.	Comparative Minimum Phenotypic F_{ST}s ($h^2 = 1$) Based on Cranial and Dental Data 36
Table 2.1.	Ethnogenetic Model of Biosocial Transformation 45
Table 3.1.	Models of Ethnogenetic Change 50
Table 5.1.	Timeline of Conflict Events during the Spanish Mission Period 95
Table 6.1.	Aggregate Pathology and Health Data from La Florida Samples 121
Table 7.1.	Significant Events in the Process of Seminole Ethnogenesis 134
Table 8.1.	Early Precontact Skeletal Samples from La Florida and Central Georgia 159

Table 8.2.	Late Precontact Skeletal Samples from La Florida and Central Georgia 160
Table 8.3.	Protohistoric (postcontact, pre-mission) Skeletal Samples from La Florida and Central Georgia 161

Preface

Edited volumes about the colonial experience in North America are very light on bioarchaeological contributions. When these are present, often only a single chapter is included, and this often deals with mortality, health, and demographic collapse. Is this because non-bioarchaeologists feel that our contributions to the field are minor or tangential? Or is it because bioarchaeologists are engaged in a rather self-contained academic pursuit? Ultimately, I do not have the answers to these questions. I do, however, hope the ideas set forth in the following chapters spark a greater acceptance and incorporation of biological datasets into general studies of history and social process as well as an appreciation among non-bioarchaeologists that we too can contribute novel perspectives on the past that complement existing, well-known archaeological and historical datasets. Yes, skeletal data have limits and it is important for bioarchaeologists to understand those limits. It is equally as important, however, for bioarchaeologists to speak broadly within the social sciences about pattern and process in order to avoid being relegated to a specialized subfield of inquiry. I hope, therefore, that this book has a wider readership than just skeletal biologists, and that the perspectives set forth in the following pages spark new ways of thinking about bioarchaeological datasets within the broader field of anthropology.

Many people contributed to this project in myriad ways, from analytical assistance to support during data collection. Here, however, I would like to draw specific attention to the one individual who sparked my interest in ethnogenetic theory, Dr. Jonathan Hill, my former department chair at Southern Illinois University. It was actually at dinner during my job interview there that Jonathan suggested I read his book, in particular the chapter by Richard Sattler on Seminole ethnogenesis. Now it is common enough for people to suggest that you read their books, but Jonathan's suggestion changed the course of my career and reaffirmed the power and importance of undirected browsing for developing problem orientations in academia. Richard Sattler, as it turned out, supported my thoughts completely about population continuity between the Spanish and Seminole periods and it was his encouragement that pushed me to proceed in this direction. E., of course, is inclusive.

Introduction

The Bioarchaeology of Ethnogenesis

> If their hands and noses were cut off they made no more account of it than if each one of them had been a Mucius Scaevola of Rome. Not one of them, for fear of death, denied that he belonged to Apalache; and when they were taken and were asked from whence they were they replied proudly: 'From whence am I? I am an Indian of Apalache.' And they gave one to understand that they would be insulted if they were thought to be of any other tribe than the Apalaches.
> —Rodrigo Ranjel, *De Soto's Expedition* (Bourne 1922: 80)

The above quotation, taken from Rodrigo Ranjel's account of the de Soto *entrada*, conveys the essence of the anthropological study of "identity." The referent group, the Apalachee, lived in the Florida panhandle and was among the most well-known and respected of Florida's indigenous populations. The strong sense of pride conveyed in this quote is still evident in the Apalachee people, as I witnessed during a ceremony commemorating the 300th anniversary of the abandonment of mission San Luis resulting in the Apalachee diaspora to lands east, west, and north, never to fully return to their native homeland. The few remaining Apalachee, Chief Gilmer Bennett and his descendants, were in attendance as a testament to their cultural heritage and connection to the landscape of the Tallahassee Red Hills. Clearly to the individual quoted by Ranjel it meant quite a lot to be Apalachee. This ethnic marker carried with it status in regional political affairs and in some way helped to define a person and a people. Manners, customs, appearance, and language all signaled this identity and defined it in relationship to other social communities that were also incorporated into Spain's colony in the late 16th century, and, like the Apalachee, fared poorly throughout the colonial era in the southeastern United States.

This book is ultimately about these people, the Southeast's indigenous communities, their struggle under Spanish rule, their lifestyles and transformations, and how they navigated the course of 16th- through 18th-century history and adapted in ways that produced a new identity, the Florida Seminole. This story of Spanish explorers, the missions that followed, English slave raids, and Creek and Seminole political machinations has been told countless times through the lens of history and archaeology. Here, I add a biological component to the saga of collapse and regeneration and present a different perspective on Florida's indigenous tribes, one that is explicitly interdisciplinary in practice with primary focus on the evolutionary signatures of the colonial experience that resulted in the formation of a new ethnic consciousness among La Florida's indigenous communities. Human biology's contribution to this narrative is the application of microevolutionary methods on archaeological human skeletal data sets to enhance regional historical narratives with nuanced interpretations that might otherwise be invisible. The indelible genetic signatures writ in the human skeleton are free from social manipulation and transcend vagaries of archaeological preservation, middle-range theoretical concerns, and textual biases. They allow us to infer what simply cannot be seen and, more important, cannot be manipulated or feigned. As one noted scholar of ethnic identity commented,

> An individual can change his name, ignore or conceal his origins, disregard or rewrite his history, adopt a different nationality, learn a new language, abandon his family's religion or convert to a new one, embrace new mores, ethics, philosophies, take on new styles of life. But there is not much he can do to change his body. (Isaacs, 1974: 30–31)

Although the historical literature has focused on demographic collapse and associated mechanical evolutionary effects,[1] models derived from a broader archaeological literature on institutional or political collapse (Parkinson, 2002; Schwartz and Nichols, 2006) can also be incorporated into this theoretical framework. When collapse occurs, people do not outright disappear. Institutions do go extinct, but the people themselves persist, and cycles of societal collapse and regeneration are, or may be, part of the normal human historical experience (Moore, 1994a, b, 2001; see also Banks, 1996: 20–24 for a historical review of Marxist and Soviet perspectives). This is the essence of the uniformitarian application of ethnogenetic theory (Moore, 1994a, b, 2001). As developed in this book, social identities also do not simply go extinct. Rather, they change and evolve as well, and produce novel and unique forms with more ephemeral connections to the past that require careful interdisciplinary reconstruction. While the names of past people may be unrecorded in the pages of history, or recorded in altered and objectified forms, biological data provide a more enduring and impartial record of experience and collective action.

In this book I attempt to reconstruct changes in social identities among Catholic communities in Spanish Florida based primarily on patterns of biological variation, supplemented extensively with archaeological and ethnohistorical data and interpreted within the broader corpus of social theory on ethnic and community identities. I draw a theoretical analogy directly from historical ethnography and the concept of an ethnogenetic "life-cycle" (*sensu* Hickerson, 1996) with its attendant diachronic focus, and draw an operational analogy from the analysis of material culture style, most evident in the historical archaeological literature (e.g., Bell, 2005; Voss, 2005). In particular, I propose that changing patterns of regional phenotypic variation can be read as a *signal* of an initial, but not requisite, phase of ethnic or community emergence or boundary redefinition in a manner similar to, but considerably more powerful than, previous archaeological inferences of activity-related material culture and architectural variation. The biology-culture divide is easily bridged in humans by consideration of the complex social processes with latent and overt identity symbolism involved in the field (in the sense of Bourdieu) of human reproduction. When couched within colonial North American Indian scholarship and ethnogenetic theory, three primary research themes emerge.

1 How were indigenous societies biologically structured in the past? How did this structure change immediately after contact and immediately following the establishment of missions in Florida and Georgia? Are changes in biological variation consistent with models of ethnogenetic transformation (separation, liminality, reintegration; Hickerson, 1996)?
2 How were populations biologically integrated across sociopolitical or linguistic boundaries? Were the records of the colonial era, with details of linguistic distributional limits, buffer zones, political landscapes, and patterns of warfare,

reflected in group mating behavior? Did these boundaries really exist biologically? In other words, did social mechanisms of intertribal integration (Albers, 1993, 1996; Moore, 1994a, b, 2001; Quinn, 1993) have a biological component that was evident in Florida's tribes? If so, was intertribal biological integration ubiquitous? If not, under what circumstances did it manifest and why?

3 How did the effects of global historical trends and processes manifest at the regional and local levels in terms of identity discourses? In other words, were the principle of human interconnectedness (Lesser, 1961), the implications of World Systems Theory (Wallerstein, 2004), and Eric Wolf's (1982) criticism of the Euro-American "billiard ball model" evident during the colonial period in La Florida? Can documented processes of identity transformation be related to changes on both local and global scales?

I approach these questions from a bioarchaeological perspective with full consideration of ethnic identity theory (Barth, 1969; Bentley, 1987; R. Cohen, 1978; Eriksen, 2002; Jones, 2002), historical ethnographic perspectives on postcolonial ethnic emergence in frontier "tribal zones"(Ferguson and Whitehead, 1992; Hill, 1996a), and a critique of biological anthropology's tendency to reify social communities as biological species and reconstruct their histories using a branching (cladistic) theoretical basis (MacEachern, 2000; Moore, 1994a, b, 2001; Terrell, 2001a, b, c; Terrell et al., 1997). What emerges, I hope, is a synthetic narrative history, an invisible history, which complements a wide array of scholarship of both regional and broadly theoretical orientations.

Bioarchaeology and the Study of the Past

In Spanish colonial Florida, a majority of the bioanthropological research has occurred under the banner of the La Florida Bioarchaeology Project directed by Clark Spencer Larsen and involving a dozen collaborators and specialists (see Larsen, 2001). Through these collective efforts much has been learned about how indigenous lifestyles changed during the contact period. Alterations in community health profiles have been the primary focus of investigation, parceled into diet, pathology and disease, and behavioral adaptations (Larsen, 2001), topics which form the core of many long-term, regionally defined bioarchaeological research programs. Community health inferences reflect the interaction of the environment with the body and are decidedly nongenetic (i.e., nonheritable) in middle-range orientation. Behaviors and life course experiences are linked to physical manifestations of skeletal morphology subject to plastic forces, whether these are micro- and macroscopic indicators of growth disruption in the dentition and long bones, changes in long bone cross-sectional geometry, or the presence of specific diseases manifest in the skeleton. The current volume complements previous bioarchaeological analyses (Larsen, 1993, 2001) by adopting an explicitly genetic perspective to examine the pattern of heritable phenotypic variation in reference to archaeological and historical models of community relationships and their transformations in the wake of contact. This approach is called biodistance analysis—the study of microevolutionary processes in past populations using skeletal or dental data (see Buikstra et al., 1990; Larsen, 1997).

Biodistance Analysis and Bioarchaeological Practice

Comparison of human skeletal morphology among populations has a long history within anthropology. Unfortunately, this history has not always produced positive contributions to humanity, as craniometry formed the core of 18th- and 19th-century "race science" (Gould, 1996; Reynolds and Lieberman, 1996). As a result, it is easy to criticize all biodistance analyses as typological at best, racist at worst. However, despite continued criticism that biodistance analysis remains typological[2] (Armelagos and Van Gerven, 2003), the field *has* clearly expanded beyond broad comparison of cranial forms within a migrationist, typological framework (Buikstra et al., 1990; Stojanowski and Buikstra, 2004, 2005), with more recent focus on intrasite (Stojanowski and Schillaci, 2006) or small-scale regional analysis (Buikstra et al., 1990; Larsen, 1997; Relethford, 2003). Although global surveys of skeletal variation persist, the current study falls within the latter category (regional analysis) and should not, therefore, be viewed in the same vein as continental or global scale analyses. The only similarity is that both types of analyses use phenotypic data to infer evolutionary processes in the past.

And this brings up a second criticism of biodistance analyses—phenotypic variation has *both* an underlying genetic component and an ontogenetic (growth, environment) component (see Houghton, 1996). The criticism in this case is that environmental effects related to growth and nutrition may mask the underlying genetic signatures that reflect population history and population structure. While it is true that phenotypic variation has both genetic and environmental components to its determination (Konigsberg,

2000), nearly all continuously varying physical and behavioral phenotypes do so as well. In addition, a number of facts support the use of skeletal data for inferring patterns of population relationships in the past. For example, when genetic distances are calculated between populations, using both phenotypic and genotypic data, both sources of information produce similar inferences about the relationships between the populations, particularly when using the dentition which forms the basis of the research presented in this book (see Adachi et al., 2003; Corruccini and Shimada, 2002; Corruccini et al., 2002; Matsumura and Nishimoto, 1996; Oota et al., 1995; Shimada et al., 2004; Shinoda and Kanai, 1999; Shinoda and Kunisada, 1994; Shinoda et al., 1998). Second, recent research on the quantitative genetics of tooth size and morphology has generally supported the strong genetic signal underlying the basis of variation in these phenotypes (Hlusko and Mahaney, 2007; Hlusko et al., 2004; Rizk et al., 2008—see also Jernvall and Jung, 2000, and Kangas et al., 2004), thereby supporting the evolutionary inferences gleaned from studies of the skeletal system.

Furthermore, use of phenotypic data such as tooth size offers a number of benefits over genotypic data when archaeological population samples are considered: (1) phenotypic analyses are nondestructive, a significant benefit when the focus of study is Native American microevolution; (2) because of analytical costs and time constraints, sample sizes for phenotypic analyses are almost always going to be larger and more representative of a population, whereas the costs and time involved in producing genotypic data sets may require a more measured sampling strategy; (3) the ability to incorporate archaeological with modern data sets affords greater time depth than that typically available for genetic datasets, advances in ancient DNA extraction technologies notwithstanding; and (4) because of this, analyses of historical processes are much better implemented by biodistance approaches because data can be analyzed in microtemporal slices. Therefore, phenotypic approaches are not only justified theoretically but they offer a number of benefits that analyses of modern and even ancient DNA lack.

In my previous book on Spanish Florida (Stojanowski, 2005a) I spent a considerable amount of effort discussing and justifying the evolutionary genetic approach based on odontometric variation. That is not the intent of the current manuscript as more pressing issues must take precedence. I stress, however, an important distinction between the analyses presented in this volume and those that are highly visible in the literature and also fall under the purview of "biodistance analysis." The analyses presented here are not about allocation or group membership, as in a forensics case that a priori assumes the existence of identifiable, distinct units (usually "races"—hence the claims of typology). This mode of analysis often reduces to typological thinking out of necessity and is subject to considerable critique (e.g., Armelagos and Van Gerven, 2003; Williams et al., 2005). Rather, the approach adopted in this work is based on changes in patterns of variation through time and space. The datasets are identical in both cases; it is the approach that differs. For example, I am not interested in determining whether a particular mission church was used by Apalachee or Timucua speakers (an allocation problem) because doing so assumes correspondence between biological and linguistic variation that may be quite spurious. Rather, I am interested in observing how patterns of variation change through time and what this may imply about the underlying structure of behaviors responsible for the observed changes. Above, I have discussed the reasons that phenotypic data are suitable for this task and may actually be preferable to genetic data in archaeological contexts. The question remains, of the choices available in the human skeleton, why teeth?

Odontometrics (tooth size) offer a number of benefits that dental morphological and craniometric and cranial nonmetric variables lack (Stojanowski, 2005a: 80–83). First, preservation of the dentition is much better than for the cranium because enamel is the hardest substance in the human body. As a result missing data are minimized and sample sizes are maximized in the use of dental metrics. Second, odontometrics are easy to record and are replicable, although inter- and intra-observer error does remain a concern (see Kieser, 1990; Stojanowski, 2001). Third, metric data are statistically tractable and amenable to population genetic approaches which assume an additive genetic, polygenic mode of expression. Discontinuous variation (binary or ordinal scale data) is much more difficult to analyze, including estimation of simple distance statistics (Edgar, 2004; Konigsberg and Buikstra, 2006; Powell, 1995), because even though the underlying genetic model is likely also polygenic and additive (Hauser and De Stefano, 1989; Scott and Turner, 1997) the phenotypic mode of expression is discretely recorded. As a result sophisticated, model-bound approaches (Relethford, 2003; Relethford and Blangero, 1990; Relethford et al., 1997) cannot be applied with discontinuous data. Fourth, dental dimensions form early in life and the crown of a tooth does not remodel after it has formed (Hillson, 1996). This is critical because the early stage of formation of adult teeth (see Stojanowski, 2001: table 5.1; Ubelaker, 1978), combined with the inability of dental enamel to remodel once it has formed, reduces the period for which the final phenotype (adult tooth size) can be affected by environmental factors such as functional loading, poor nutrition, or maternal diet and health (as opposed to something like the human face, which remodels and responds to mechanical stresses throughout an individual's lifetime). Because of this lack of plasticity, odontometrics have higher narrow-sense heritability estimates than craniometrics (reviewed in Stojanowski, 2001, 2005d)[3]—not because teeth are "more genetic" than craniofacial dimensions but because the environmental variance is lower for teeth as a result of the early age of phenotypic formation. Environmental variation may be further minimized by using polar teeth only (Garn et al. 1965a, b, 1967a, 1968; Garn et al., 1967b).[4] Fifth, mesiodistal and buccolingual measurements are located at different positions on the tooth crown for each tooth type. This maximizes genotypic representation and distributes the potential confounding effects of missing data due to the wearing of the crown surface throughout the dental arcade. Finally, odontometric data are less subject to cultural (such as intentional cranial modification) and post-depositional (warping and crushing due to soil compaction) modification than craniometrics, also reducing the impact of missing data (see figure 0.1).

These aspects of comparative odontometric research have previously been developed in greater detail and specifically related to microevolutionary models incorporating gene flow and genetic drift in Stojanowski (2005a). As with my prior consideration of these data, which consist of mesiodis-

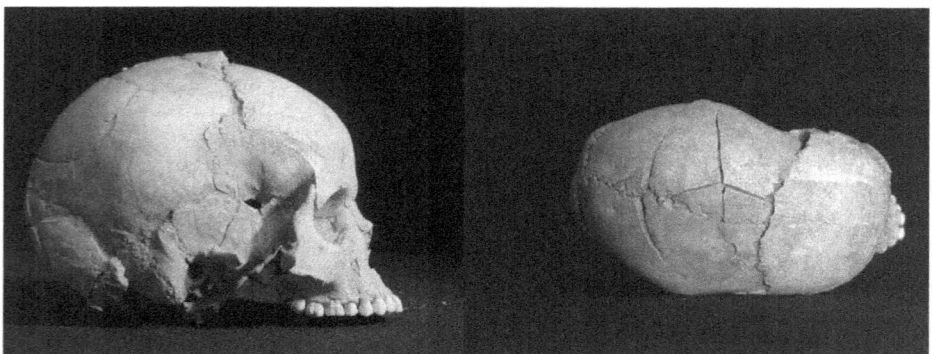

Figure 0.1. Two views of a cranium showing the effects of distortion on cranial shape. From the right lateral view the crushing of the vault is minimally apparent. However, in superior view it is clear that craniometric analysis could not include this specimen. Note that the teeth are perfect. Please note that this is not a Native American skull.

tal and buccolingual polar tooth dimensions for the maxilla and mandible (Stojanowski, 2001, 2003a, b; 2004; 2005a, b, c, d), the effects of mutation and natural selection on patterns of phenotypic variation are considered minimal in comparison to the expected effects of gene flow and genetic drift on population structure. Almost all *micro*evolutionary studies exclude mutation as a significant cause of phenotypic variation because of the short time periods under consideration. Mutation rates are too slow to be significant when only a few generations are sampled, as is the case here. Admittedly, it is more difficult to assume natural selection was not operative. However, given the short time period considered; the known relationship between mortality, morbidity, and epidemic disease; and the unlikely association between these factors and tooth size, it is not unreasonable to assume that selection effects were also minimal. And, despite the use of tooth size as a proxy for body size in paleontological contexts, several studies have found low correlations between tooth and body size in recent human populations (Filipsson and Goldson, 1963; Garn et al., 1968; Henderson and Corruccini, 1976; Lease and Harris, 2001) such that secular trends during periods of nutritional transitions may not manifest in the dentition as a scaling effect (large body = large teeth). I continue to see little reason that selection would have been operating on tooth size, given its poor correlation with general body size, during a short time period in which a host of epidemic agents were affecting the New World colonies (Dobyns, 1983; Ramenofsky, 1987) and people were dying primarily because of the violence, morbidity, and privation that characterized the frontier zones of expanding European colonies. Therefore, microevolutionary changes in the dentition are reflective of changes in population size (genetic drift) and social identity discourses that affect the flow of people within and across mate exchange networks (gene flow). By statistically controlling for the former, one can directly access information on the latter.

Structure of the Manuscript

I have previously published parts of this work in various places but I have never been afforded the opportunity in a professional journal to fully develop the model and contextual basis for this interpretation of colonial ethnogenesis in the southeastern United States. Here I attempt to do so. The goal is to compile relevant historical and archaeological data and combine these with bioarchaeological and evolutionary genetic analyses to interpret ethnographic models of colonial ethnogenesis. To accomplish this, I adopt a nonstandard manuscript structure in which the evolutionary genetic analyses (the results, if you will) are presented in the first chapter. These are the most central elements upon which the rest of the book is based and it is important that the biological details are clearly delineated from the outset. The middle third of the manuscript then attempts a more novel interpretation of these data with reference to social theory, beginning with chapters 2 and 3 in which I explicitly link the biological and cultural realms of the human experience and develop the theoretical interpretive framework of ethnogenetic theory. After I establish the critical connection between biology, culture, and the material realm of the archaeological past, I then reinterpret the quantitative genetic data in reference to a life-cycle model of ethnogenetic change as developed in the historical ethnographic literature. Chapters 4, 5, and 6 consider archaeological and historical data that support the biological evidence for the emergence of a panethnic consciousness among La Florida's Christian communities. These materials derive from the historical ethnographic and archaeological literatures, drawing heavily on social theory related to ethnic identity emergence and transformation, practice and habitus scholarship as applied to notions of community or ethnic consciousness, and historical ethnographic concepts of tribalization and the concept of the "tribal zone."

Although indigenous ethnogenetic transformation has not often been discussed in past research on Spanish colonial Florida, the interpretation may not come as a surprise. Ethnogenesis was a common occurrence among Native American communities during the colonial period. However, this is not just an issue relegated to "academic interest" because these peoples were displaced during the 18th century and no longer "exist" in Florida as distinct ethnic or tribal identities. Rather, I propose that this interpretation of indigenous history provides a novel biosocial backdrop against which Seminole ethnogenesis (which occurred during the 18th century) should be considered. In chapter 7, I explore the migration rationale for proto-Seminole emigrations from central Georgia back into the old mission fields of La Florida. In particular, I propose (as have others) that the proto-Seminole communities were not just disaffected Creeks lured into Florida by the promise of Spanish trading or a rich resource base. The proto-Seminole were in many cases the very same ethnic groups that had lived in Florida during the 17th century, and as such, Seminole ethnogenesis should be viewed as a conscious return to ancestral lands by peoples whose identities were replaced in the colonial record but preserved in the physical remains of their bodies. The biosocial basis of this later history, however, was forged during the 17th century and revealed by the analyses of phenotypic variability presented in chapter 1. As such, there is a direct biological *and* social connection between the Apalachees, Guales, and Timucuas of the 16th and 17th centuries and Creeks and Seminoles of the 18th century. To deny this is to ignore the complex patterns of biological integration that innervated communities throughout the southeastern United States, both before and after contact. And in chapter 8 I return to the biological data once again and consider the extent of community integration among *precontact* populations in Florida and Georgia using a more comprehensive sampling design that includes the Georgia interior. This final substantive chapter approaches the issue of identity in a slightly different manner. In evaluating whether the Seminole are "true" Florida Indians, an inherent bias in the framework of discourse is highlighted. To ask this question in this way assumes that indigenous populations in Florida were somehow distinct from others in the Southeast, thereby allowing one to draw a false dichotomy between "true" Florida Indians and foreigners. Data presented in chapter 8 indicate that this was never the case, although the structure of relationships did change through time in meaningful ways. This circular, biohistorical narrative is summarized in the concluding chapter (chapter 9) and some additional areas of impact are identified.

One *Genetic Landscapes
of Spanish Colonial Florida*

This chapter details the microevolutionary processes that were operative during the late precontact through First Spanish periods in colonial Florida and Georgia. The Native American communities subject to Catholic conversion during this time period are the focus of analysis, and the interpretive framework invokes consideration of patterns of community integration. Two key temporal transitions are considered. The first targets the late precontact to early postcontact interval, documenting changes in population structure and community organization during the earliest phases of missionization (ca. 1600–1650). The stated date of 1600 is somewhat arbitrarily determined by the archaeological sampling as late 16th-century missions have not been identified and systematically excavated as of this writing. Nonetheless, the inferences about this first transition are analogous to those which consider broadly conceived consequences of contact. The second transition targets the early through late postcontact interval, centered on the mid-17th century (ca. 1650–1700). The end date is approximated by the destruction of the missions during the first decade of the 18th century. To understand the choice of the mid-century date (1650) dividing the early and late mission periods, a better understanding of the timeline of significant events in La Florida's history is needed. Although the date of 1650 is, in part, also reflective of the constraints of archaeological sampling, there was a significant demographic and social transition that occurred around the mid-17th century. I turn to these details next, followed by a formal presentation of the microevolutionary analyses including the sampling design, methodology, and analytical results.

A Brief History of Spanish Colonial Florida

Spain's interest in Florida began with a series of 16th-century *entradas* into North America postdating the successful colonization ventures among the Aztecs and Incas (Milanich, 1990). These earliest *entradas* by de León (1513, 1521), Narváez (1528), and de Soto (1539), among others, were unsuccessful colonization efforts, making Spain's claim to North America tenuous. Spurred by French Huguenot activities along the Atlantic coast, the task of permanent colonization would fall on Pedro Menéndez de Avilés who established St. Augustine in 1565. As part of his obligations involved ideological conversion, friars were a component of the earliest colonization efforts, reflective of the shift in Spanish colonial policies toward pacification of indigenous populations.[1] By 1573 a series of Franciscan missions had been established around St. Augustine and along the Florida and Georgia coast. The peak period of expansion along the coast occurred from 1595 through 1620 when populations living along the Georgia coast (Guale province) and along the Atlantic coast of Florida and Georgia, including the St. Johns River drainage (eastern Timucua province), were converted (Milanich, 2004). By the 1620s, missions had been established throughout north-central Florida (western Timucua) and by 1633 among the Apalachee in the eastern Florida panhandle (Hann, 1988). Although missionary efforts were attempted beyond these boundaries (Hann, 1991, 1993b; Thomas, 1988), they were short-lived. At the height of missionization circa 1650 several dozen missionaries served communities throughout north-central Florida into the panhandle and northward along the Florida and Georgia coast (figure 1.1).

These communities were organized into three provinces by the Spanish: Apalachee, Guale, and Timucua (see locations of these provinces in figure 1.2). Such nomenclature implemented colonial administration but minimized the true degree of cultural variation that existed at first contact (see entries in McEwan, 2000a; Milanich and Proctor, 1978). Both Apalachee and Guale provinces were relatively internally culturally homogenous.[2] The Apalachee were a Muskogean-speaking chiefdom located between the Aucilla and Ochlockonee Rivers in the Florida panhandle. They were dedicated, sedentary maize agriculturalists living at fairly high population densities (Hann, 1988; McEwan 2000b; Scarry, 1994). The Guale lived along a narrow strand of the Georgia coast between the Altamaha and Ogeechee Rivers.[3] They were less sedentary than the Apalachee and relied less on

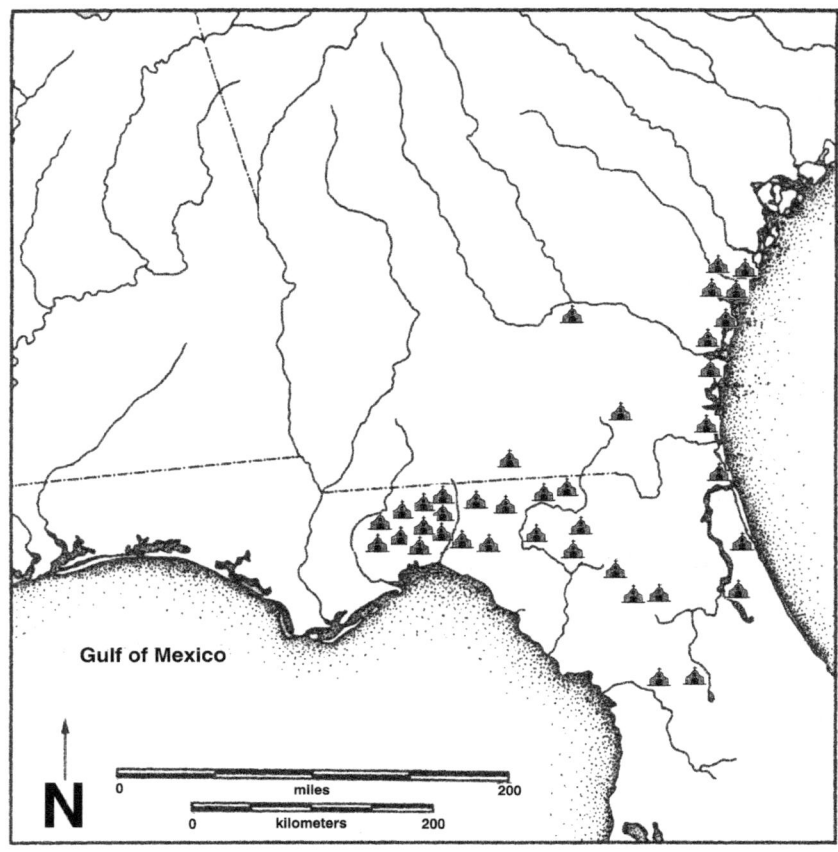

Figure 1.1. Map of La Florida indicating approximate locations of mission *doctrinas* circa 1650 (modified after Larsen 2001, fig. 2.1).

maize agriculture and more on marine and estuarine resources, a subsistence strategy shared with their Timucua neighbors[4] (Jones, 1978; Larson, 1978; Saunders, 2000a; Worth 2004). It is believed that the Guale also spoke a Muskogean language.[5] On the other hand, Timucua was a polyglot province with no real political meaning in prehistory.[6] The Timucua chiefdoms were never united into a single confederacy but rather represented a large linguistic category comprising 25–50 distinct chiefdoms,[7] speaking a dozen dialects[8] with considerable variability in settlement structure, subsistence strategies, and material cultural inventories (Deagan, 1978a; Ehrmann, 1940; Milanich, 1978, 1996, 2004; Milanich and Sturtevant, 1972). Most agree that the Timucua languages were not Muskogean in macroaffiliation.[9] Timucua languages were not intelligible to their Muskogee-speaking neighbors.[10]

The years 1632 through 1674 have been labeled the "Golden Age" of the Spanish missions because this period witnessed the peak expansion of missionary activity throughout the colony (Geiger, 1937). However, as noted by Spellman (1965), and what has become abundantly clear from decades of historical and bioarchaeological research, this optimistic declaration is something of a misnomer. Economically the colony was under constant duress (Bolton, 1917; Bushnell, 1981, 1994; Geiger, 1937; Lyon, 1990), there was tension between religious and secular segments of the Hispanic population (Arnade, 1960; Gannon, 1965; Matter, 1972), and attempts to expand the missions were hindered by poor transportation and communication inherent

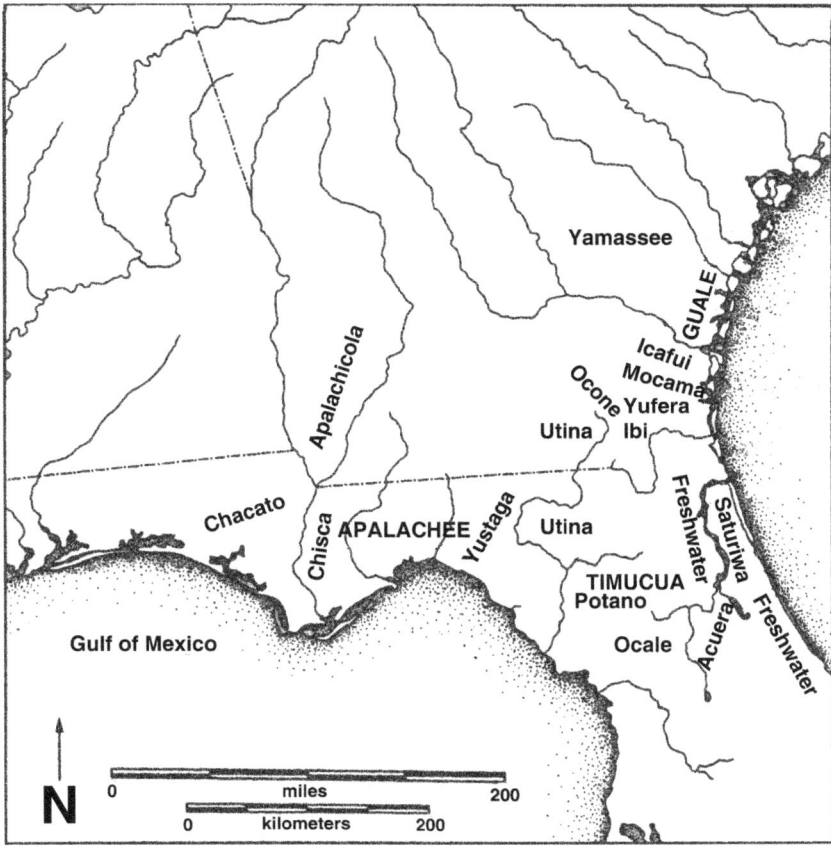

Figure 1.2. Cultural groups incorporated into the Spanish mission system (modified after Larsen 2001, fig. 2.1). Locations of specific ethnic or tribal populations are based on data presented by Hann 1996a (p. 2, map 1).

to Florida geography (Thomas, 1990a). That the mission, by definition and design a frontier institution, was never supplanted by secular, private governance is also telling (Bolton, 1917; Thomas, 1988) and reflects the inability of the missionaries to effect a Spanish residence pattern among the converted populations (Bushnell, 1990; Deagan, 1990a).

Life for members of the Christian mission communities did not improve after conversion to Catholicism. Native Americans suffered frequent epidemics (Bushnell, 1978; Deagan, 1973, 1990b; Dobyns, 1983, 1991; Hutchinson and Larsen, 1988; Jones, 1978; Swanton, 1922) and the onerous demands of the *repartimiento* system (see Bushnell, 1979: 5, 2006; Geiger, 1937: 200; Spellman, 1965: 361, 366; Worth, 1998b: 21 for specific examples). Declining health conditions exacerbated demographic collapse resulting in population size decline (Deagan, 1978a; Dobyns, 1983; Hann, 1988, 1996a; Milanich, 1978, 1996, 1999; Worth, 1995, 1998b), increased fugitivism (Worth, 1998b), and frequent revolts which were eventually supplanted by slave raids from populations serving the needs of traders affiliated with England's southern colonies (Covington, 1967; Crane, 1956; Hahn, 2004; Worth, 1995). Population losses, whether due to epidemics, labor abuses, slave raiding, or fugitivism, stressed the already weakened system. And after nearly 150 years of struggle and hardship, the Spanish missions were destroyed during a series of raids by English-serving Apalachicola/Uchise/Creek Indians in the late 17th and early 18th centuries (Arnade, 1959; Boyd et al., 1951). By 1706, St. Augustine was all that remained of Spain's La Florida colony (see table 1.1).

Previous Bioarchaeological Research

Bioarchaeological contributions have provided new perspectives on changes in lifestyle experienced by Christian native communities in Florida and Georgia during the intervening years of the 17th century. Bioarchaeological research has confirmed that maize became an ever more important component of the diet after missionization, an inference based on analysis of caries frequencies which increased through time (Larsen et al., 1991, 2002), light isotope analyses of carbon and nitrogen indicating an increase in maize consumption at the expense of marine resources (Hutchinson et al., 1998, 2000; Larsen et al., 1992, 2002; Schoeninger et al., 1990), incisor and molar enamel microwear indicating dietary homogenization after missionization (Teaford et al., 2001), and trace elements analysis suggesting the reduction of wild plant foods in the diet (Ezzo et al., 1995). That diet and morbidity are syner-

Table 1.1. Important dates in La Florida history

Date	Event
1513	Ponce de León founds La Florida
1526	de Ayllón founds San Miguel de Gualdape
1528	Pánfilo de Narváez entrada
1539–1543	de Soto entrada
1562	Charlesfort founded
1563	Charlesfort abandoned
1563	Fort Caroline founded
1565	St. Augustine founded
1566	First Jesuit missionaries arrive
1570	Menéndez relieved of Florida overlordship
1572	Jesuits leave La Florida
1573	Franciscans arrive
1597	Juanillo revolt in Guale
1608	Peace negotiated between Apalachee and Timucua
1633	Apalachee missions established
1647	Apalachee revolt
1656	Timucua revolt
1659	Westo arrive in Southeast
1668	St. Augustine attacked by Robert Searle
1670	Charles Town founded
1672	Castillo de San Marcos begins construction
1687	Castillo essentially complete
1702	Moore attack on Guale coast/St. Augustine
1704	Moore attack on Apalachee and Timucua

gistic is well known (Larsen, 1997), and pathological signatures reflective of poor community health also increase after missionization. Changing enamel hypoplastic defect frequencies indicated an increasingly stress-filled environment, often assumed to be dietary in nature but somewhat complicated by the complex etiology of these markers of stress in the dentition (Hutchinson, 1986; Hutchinson and Larsen, 1988, 1990, 2001; Larsen and Hutchinson, 1992; Storey, 1986). Accentuated striae of Retzius frequencies also reflected the effects of stress, particularly in the earliest years of life (Simpson, 2001; Simpson et al., 1990). Frequencies of cribra orbitalia and porotic hyperostosis peaked during the mission period (Larsen and Sering, 2000; Schultz et al., 2001) and the prevalence of generalized bone infections of the periosteum signaled an increase in injury-producing behaviors (Larsen and Harn, 1994). The burdens of the *repartimiento* labor system, and even quotidian life *bajo campana*, were signaled by increased rates of arthritic modification of the joints (Griffin and Larsen, 1989; Larsen, Ruff, Griffin, 1996), homogenization of male and female upper limb bone robusticity (Fresia and Ruff, 1987; Fresia et al., 1990), and changes in overall activity patterns and body size (Larsen and Ruff, 1994; Ruff and Larsen, 1990, 2001).

The above summary demonstrates that the signatures of declining health were clear. Equally as critical for understanding the mission period, however, is an assessment of the evolutionary mechanisms operating concurrently within and among mission communities. Changes in health must be interpreted within the context of population structure and demography. Migration or gene flow and population size are two components of population demography, both of which can be examined using microevolutionary models. I turn to these details next, although I note the purpose here is not to present a demographic backdrop to the health transition. Rather, the purpose of this chapter is to define the evolutionary processes operative during the 17th century.

Model-bound Population Genetic Analysis

The last 25 years have witnessed significant advances in methods for analyzing patterns of phenotypic variation in extant and archaeological populations. Traditional approaches of population-level morphological comparison were popularized by W. W. Howells (Howells, 1989, 1995) and developed concurrently with advances in statistical computing and the incorporation of multivariate statistics into biological anthropology. Excellent summaries of these anthropological research methods have been presented by Pietrusewsky (2000) and Kieser (1990) and a more general, yet accessible, discussion by Manly (1994). As no specific population genetic parameters are estimated these multivariate statistical approaches are "model-free" (Relethford and Lees, 1982) and rely on "statistical analogies to specific population structure models" (Relethford and Blangero, 1990: 6).

However, beginning in the 1980s researchers became interested in modifying allele-based population genetic analyses for use with phenotypic data of continuous scale. More specifically, the Harpending and Jenkins (1973)

R-matrix model (Relethford and Blangero, 1990; Williams-Blangero, 1989a, b; Williams-Blangero and Blangero, 1989) and the Harpending and Ward (1982) model for detecting differential extra-local gene flow were modified for use with continuous, quantitative data (Relethford and Blangero, 1990). These methods were subsequently fine-tuned to correct for statistical sampling bias and expected drift distances, and standard errors were defined for output statistics, allowing calculation of p-values (Relethford, 1991a, 1996; Relethford et al., 1997). The most current summary of phenotypic R-matrix analysis is presented in Relethford (2003). The methods were codified in the statistical software package RMET which performs the calculations for multivariate datasets (Relethford, 2003; Relethford et al., 1997).

These methodological advancements are particularly noteworthy because they position analysis of morphological microevolution within a population genetic framework and, under certain circumstances, are model-bound in that parameters from theoretical models are directly estimated from the data (Relethford and Blangero, 1990; Relethford and Lees, 1982). In addition, more formal methods provide modeling capabilities not possible with multivariate statistical analogues. For example, intersample distances can be corrected for intervariable covariation and different modeling scenarios can be generated that account for variations in effective population size (genetic drift) and differences in trait heritability.[11] Measures of regional genetic variation such as F_{ST} provide a summary measure of between-population genetic variation that is similarly corrected for the effects of trait heritability, differential effective population size, and sampling bias.

These inferential benefits are not without significant assumptions, however (Relethford and Lees, 1982; Williams-Blangero and Blangero, 1989). For example, phenotypic R-matrix analysis assumes that the traits are selectively neutral; follow an equal and additive effects model of trait expression; share a common, positive heritability across traits and populations that is invariant through time; and are not affected by dominance, epistasis, major gene effects, or genotype-environment interaction effects (Blangero, 1990; Relethford and Blangero, 1990; Williams-Blangero, 1989a, b; Williams-Blangero and Blangero, 1989). While the details are beyond the scope here, odontometric variation reasonably conforms to these expectations (relevant literatures are discussed in Kieser, 1990; Stojanowski, 2001).

Regardless of these strict assumptions, the analytical results of model-bound methods are extremely powerful and have completely supplanted multivariate statistical approaches for landmark data (see, for example,

González-José et al., 2001; Nystrom, 2006; Powell and Neves, 1999; Relethford, 2001; Scherer, 2007; Schillaci, 2003; Schillaci and Stojanowski, 2005; Sparks and Jantz, 2002; Steadman, 1998, 2001; Stefan, 1999; Tatarek and Sciulli, 2000; Varela and Cocilovo, 2002). Two analytical results are of primary interest: interpopulation genetic distances and estimates of regional genetic microdifferentiation, a parameter called F_{ST}. Rather than focus on implementation (the mechanics), it is more critical to outline the manner in which these two statistics are meaningfully interpreted.

What Do Genetic Distances between Populations Really Mean?

Genetic distances can be directly estimated from an R-matrix (Relethford, 2003, equation 3.2) and, assuming complete heritability ($h^2 = 1$), represent *minimum* estimates (Williams-Blangero and Blangero, 1989). These distances are, therefore, conservative measures of population divergence and are comparable to distance ordinations produced using traditional statistical measures. The concept of a distance is relatively straightforward. Populations that exchange mates become more genetically similar. Allele frequencies converge toward an equilibrium value and between-population genetic variability decreases. This manifests phenotypically as similar mean values for continuous measurements or similar frequencies of trait expression for discontinuous traits (figure 1.3). By comparing genetic distances among populations for different time periods one can infer how patterns of biological integration changed through time. In figure 1.4, for example, I present a distance ordination for three populations for two time periods, represented by circular (time = 1) and square (time = 2) icons. The interpretation of the distances in this figure is straightforward. Populations A and B become more similar through time, while population C diverges from A and B. If the effects of genetic drift and differential effective population size are mitigated, either through statistical scaling (Relethford, 1991a, 1996, 2003) or the assumption that population sizes were roughly equal, then genetic distances represent patterns of similarity due to mate exchange and migration. If two populations increase in similarity, this suggests increasing gene flow among them, and vice versa.

Distances extracted from an R-matrix offer several benefits not possible with model-free multivariate statistical approaches. Distances can be calculated using varying estimates of narrow-sense heritability (from 0 to 1 for all traits aggregately) to gauge how interpopulation distances change as propor-

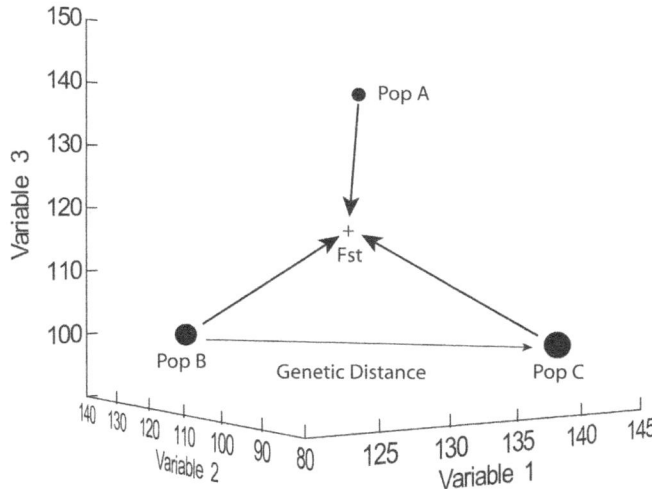

Figure 1.3. Hypothetical distance ordination demonstrating the conceptual basis of the model-bound analyses used in this chapter. Three populations are represented (A, B, C) in multivariate space, and the mathematical basis of genetic distances and phenotypic F_{ST} is indicated visually. The cross in the center of the plot indicates the position of the regional centroid or average of each of the three variables represented in this figure.

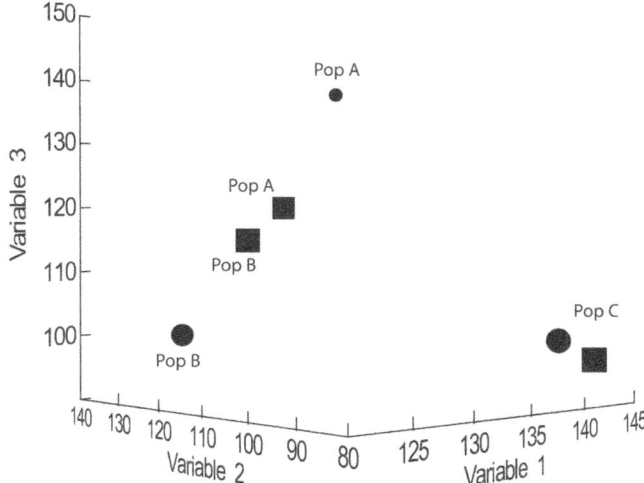

Figure 1.4. Hypothetical distance ordination for three populations (A, B, C), each measured at two distinct time periods. The circles represent the earlier samples (time = 1), and the squares represent the later samples (time = 2). As indicated, populations A and B are more similar to each other at time = 2, reflective of increased biological interaction between these populations, while population C is more divergent at time = 2.

tional additive genetic variance decreases due to inbreeding or natural selection, or as environmental variation increases among individuals within a population. Although it is unfortunate that the same heritability must be used for all populations and all traits (Carson, 2006), it is not possible to estimate sample-specific heritabilities for archaeological populations and one cannot simply transfer the heritabilities reported for specific measurements from one population to another. Therefore, although it could be more informative to consider synchronic patterns of genetic distances under an assumption of differential narrow-sense heritabilities for individual measurements and populations, implementing such an approach would be empirically unjustifiable as it requires detailed information on pedigrees that are rarely available for archaeological samples. In addition, quantitative genetic R-matrix methods implement the inclusion of important evolutionary parameters such as differential effective population size. This is a critical point because if population size varied significantly among populations either synchronically or diachronically the resulting genetic distance matrix will be highly sensitive to these differences. This modeling capability adjusts for the effects of genetic drift (Relethford, 1991a, 1996; Relethford et al., 1997), a rescaling process that theoretically reveals population structure due solely to migration. Of course, population size did decline throughout the mission period and previous research suggests the timing of demographic collapse was not homogenous throughout the Spanish provinces (Stojanowski, 2005a).

Interpretation of F_{ST} Statistics

In addition to genetic distances, quantitative R-matrix analysis produces an estimate of regional genetic variability similar to Wright's F_{ST}. (Williams-Blangero, 1989b; Williams-Blangero and Blangero, 1989). F_{ST} is a single value that summarizes within-population and among-population variation simultaneously. The statistic can be compared both intraregionally through time and with estimates reported in other studies to determine, in a comparative sense, how well cultural variability and linguistic variability correlate with genetic variability. As calculated, the statistic represents the average weighted distance of each sample from the regional centroid (the mean of means) and therefore serves as an *estimate of overall diversity within a mating network* (see figure 1.3). If samples are genetically similar they are all closely approximated by the mean. Likewise dissimilar populations are poorly represented by the mean centroid.

More important, however, is how one interprets changes in F_{ST} through time. Random migration among populations reduces overall variability, while a decrease in population size, a decrease in migration, or nonrandom migration, such as kin- or clan-structured migration, increases regional genetic variance and F_{ST} (Relethford, 1991b; Williams-Blangero, 1989a, b). Increasing population size has a more complex effect upon patterns of genetic variability (Relethford, 1991b) with predicted changes highly dependent on demographic parameters. Fortunately, we know population size did not increase after contact, which simplifies interpretation: a diachronic increase in F_{ST} suggests decreasing population size with no change in migration rates *or* a reduction in the rate of migration among populations. A diachronic decrease in F_{ST} suggests an increase in the rate of migration among populations.

Nonetheless, it is important to consider *both* the regional level of variability (F_{ST}) and the pattern of genetic distances (population structure) when inferring the microevolutionary processes responsible for a given population structure. For example, in figure 1.4, F_{ST} decreased between time periods (from circle to square icons) because the rate of change between populations A and B was greater than the divergence demonstrated by population C. All three populations are more similar to than different from one another overall as measured by F_{ST}; however, this statistic masks some of the underlying processes responsible for the changes. In the example presented in figure 1.4 all of the decrease in F_{ST} is attributable to increased biological interaction between populations A and B.

While the ability to estimate F_{ST} directly from phenotypic data is a significant improvement for prehistoric biodistance research, the point estimate itself has limited meaning except in a comparative context. It cannot be directly equated with a specific rate of migration among populations or in a demographic model; an F_{ST} of .05 or .15 means little when decontextualized. Its power is strictly comparative. Although F_{ST} estimates are directly comparable, determining whether a regional mating network is more or less integrated than others requires normalizing the estimates, minimally, by the following:

1 The size of the area sampled. Larger sampling areas encompass more diverse populations, which increases F_{ST}.
2 The temporal variation in the study samples. The greater the time averaging within a cemetery the more variable that sample appears when measured skeletally. This increases F_{ST}.

3   The number of populations sampled. As the number of samples increases the measured degree of genetic diversity may also increase.
4   The type of data used to estimate F_{ST}. Different phenotypic traits may produce divergent evolutionary genetic signatures due to distinct genetic histories.
5   The heritability of the traits. This is easily controlled by assuming maximum heritability, that is, $h^2 = 1$.

Many of these concerns are obviated when F_{ST} is calculated for the same series of populations as defined geographically using well-controlled archaeological cemeteries with the same battery of phenotypic traits with a constant heritability (e.g., Stojanowski, 2004, 2005c). Because this last point is so critical to the success of the analyses, it is important that the sampling design take all of these into consideration, a topic I turn to next.

Sampling Design

The last half century has witnessed considerable growth in Spanish mission studies in the southeastern United States, and several archaeological research programs were initiated that produced a fairly extensive bioarchaeological database (for example, Jones and Shapiro, 1990). Although skeletal remains were recovered from at least fourteen specific missions (see Larsen, 1993: table 12.1), only seven of these produced samples of appropriate size for evolutionary analysis, or are currently available for analysis. Four date to the early mission period (ca. 1600–1650) and three date to the late mission period (ca. 1650–1700). Summary information for each sample is presented in table 1.2 with locations indicated in figure 1.5.

Table 1.2. Summary information on mission period samples in La Florida

Site	Province	Dates	Period
San Pedro y San Pablo de Patale	Apalachee	1633–1650	Early Mission
San Luis de Talimali	Apalachee	1656–1704	Late Mission
Santa Catalina de Guale	Guale	1608–1680	Early Mission
Santa Catalina de Guale-Santa María	Guale	1686–1702	Late Mission
Ossuary at Santa Catalina	Timucua?	?1560–?1686	Early Mission
Santa María de los Yamassee	Timucua/Yamassee	?1650–?1683	Late Mission
San Martín de Timucua (Fig Springs)	Timucua	1608–1656	Early Mission

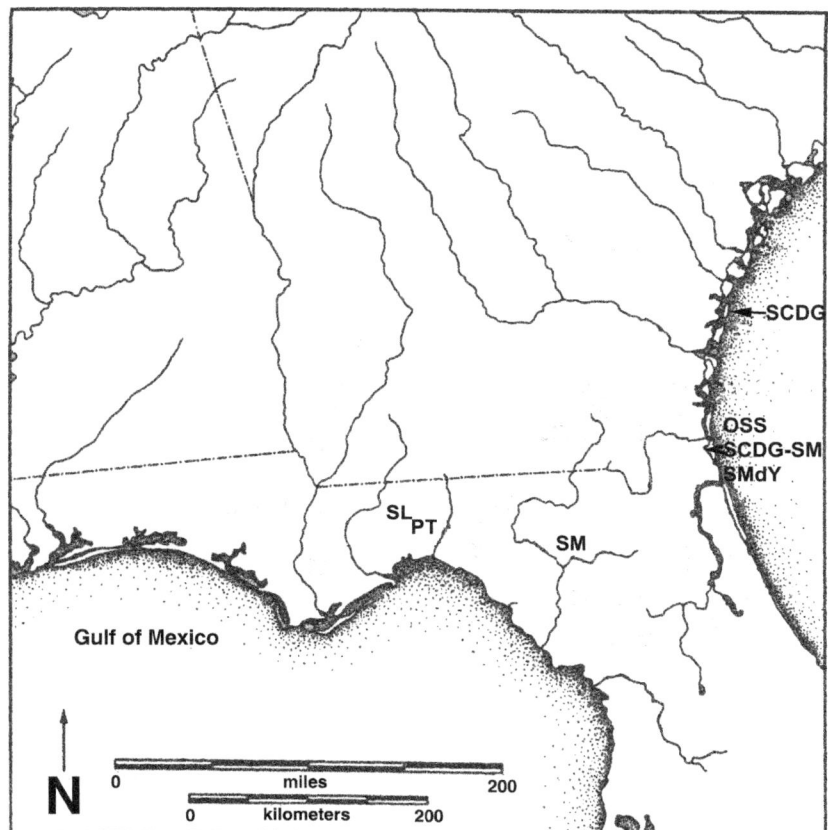

Figure 1.5. Map of Florida and Georgia with mission locations indicated. OSS = Ossuary at Santa Catalina de Guale de Santa María, PT = San Pedro y San Pablo de Patale, SM = San Martín de Timucua, SCDG = Santa Catalina de Guale, SL = San Luis de Talimali, SCDG-SM = Santa Catalina de Guale de Santa María, SMdY = Santa María de los Yamassee.

Excavations in Apalachee province produced two samples that represent nonoverlapping periods of interment that completely encompass the period of missionary activity (A.D. 1633–1704). The early mission period is represented by San Pedro y San Pablo de Patale, which dates from circa 1633 to 1650 where the *terminus ante quem* is established by the *majolica* inventory (Jones et al., 1991; Marrinan, 1993; Jones et al., 1991: 73). Late mission period Apalachee are represented by burials from San Luis de Talimali, which dates from circa 1650 to 1704 and was home to close to 1,500 Apalachee who lived alongside several hundred Spaniards during the last half of the 17th century (Boyd et al., 1951; Hann, 1988; McEwan 1991a, b, 1992, 1993a, b, 2000b, 2001).

Moving east from Apalachee into the vast Timucua interior, it is surprising that there are few postcontact samples representative of the western Timucua. Larsen (1993) notes only four[12] mission period skeletal samples (Santa Fé de Toloca, San Juan del Puerto, San Pedro y San Pablo de Potohiriba, and San Martín de Timucua). Of these, only San Martín de Timucua produced a sample of sufficient size and with the appropriate degree of skeletal preservation for inclusion in this study. The site is located in the Utina district and dates from circa 1608 to 1656; the *terminus ante quem* is established by the Timucua rebellion that destroyed many missions within the Florida interior (Weisman, 1993; Worth, 1998b).

As with Apalachee, the Guale are also represented by two mission samples that do not overlap temporally. However, unlike Apalachee these samples are actually known to represent the same congregation and therefore reflect a formal biological lineage. Santa Catalina de Guale is the earlier of the samples and is located on St. Catherines Island, Georgia. Because the Guale missions were destroyed during the 1597 uprising that affected the coastal provinces (Lanning, 1935), it is believed that the mission sample represented here has an initial date of 1608 (Thomas, 1990b). Santa Catalina de Guale was abandoned in the 1680s (Larsen, 1990) when the congregation moved farther south and was combined with other *doctrina* populations in the wake of the demographic collapse and slave raiding that caused high mortality along the coast (Worth, 1995). After abandoning St. Catherines Island, the Guale population of this mission eventually resettled farther south to be closer to St. Augustine (Worth, 1995). The location of this new mission was on Santa María (or Amelia) Island and the mission is historically documented as Santa Catalina de Guale de Santa María (Hardin, 1986; Larsen, 1993). This site was used for a brief period, from 1686 to 1702 (Larsen, 1993; Saunders, 1993).

The final two postcontact samples are also located on Amelia Island. The first is an ossuary that was located under the northwest corner of the Santa Catalina de Guale de Santa María church. Bioarchaeological analysis of this sample suggested that the remains were formerly stored in a charnel structure (Larsen, 1993). Because one burial from the Santa María main church intruded into the ossuary, and because no ossuary remains were found in the church fill, the ossuary had to have been placed there before the Santa María church was constructed. Therefore, the latest date for this site is 1686, although it likely predates Santa María considerably (Larsen, 1993). In addition, two coffin burials were found underneath the main ossuary accumulation, thus indicating a postcontact date of burial. Beyond these broad ranges it is impossible to know the exact temporal position of this sample. Although located underneath a church attributed to ethnic Guale, the sample is located on an island that traditionally fell within Timucua territory, and Larsen (1993) considered the ossuary remains ethnically Timucuan.

A second church structure was located 40 meters south of the main cemetery of Santa Catalina de Guale de Santa María (Saunders, 1993), which is believed to be the mission of Santa María de los Yamassee. There is limited information in the historic record about the inhabitants of the Santa María mission, and the site's exact date of occupation is uncertain. The mission is not on the 1655 mission list but is listed on the 1675 Calderon enumeration as a residence for 40 non-Christians (Saunders, 1993: 36). If the site actually is Santa María de los Yamassee there is no record of its founding date but it likely postdates mid-century but predates the founding of Santa Catalina de Guale de Santa María. Again, if this is a Yamassee mission, it must have been abandoned by 1683 when the entire Yamassee population of La Florida fled the Spanish territories for the Georgia interior (Bushnell, 1986: 5).

The precontact database must be populated with samples that provide the most accurate measure of regional genetic diversity and population relationships *immediately* preceding the contact period. In addition, the geographic distribution of the precontact samples should closely follow that of the early and late mission period samples. With these caveats in mind, pre-mission period samples were selected from three broad regions. The precontact Guale are well represented by samples from the same geographical area as the Guale of the historic period. Seven samples were selected that date from around A.D. 1200–1550 (table 1.3). All represent postagricultural populations living along the Georgia coast, and approximately 159 individuals are represented in total. The precontact Apalachee are also well represented geographically by suitable precontact analogues. Although four different samples were included (table 1.3), which date from approximately A.D. 1200–1500, the sample sizes are small, with 26 individuals represented. Whereas the Apalachee and Guale were well represented geographically, the interior Timucua were not. To provide a nearest suitable analogue for the San Martín de Timucua mission, data were used from west-central Florida, technically associated with the Tocobaga ethnic group. Three samples are included (table 1.3) representing 70 individuals.

Table 1.3. Late precontact skeletal samples from La Florida

Site	Province	Location	Date (A.D.)	Reference
Little Pine Island	Guale	Sapelo River/Coast	1200–1300	Larsen, 1982
Norman Md (9McI64)	Guale	Sapelo River/Coast	1200–1300	Larsen, 1982
7 Mile Bend Md (9Bry6)	Guale	Ogeechee River/Coast	1200–1550	Larsen, 1982
Kent Md (9Gn51)	Guale	St. Simons Island	1300–1550	Larsen, 1982
Lewis Creek Mds (9McI88)	Guale	Altamaha River/Coast	1200–1300	Larsen, 1982
Irene Mound Site (9Ch1)	Guale	Savannah River/Coast	1300–1550	Hulse, 1941
South End Md (9Li3)	Guale	St. Catherines Island	1200–1550	Larsen, 2002
Tierra Verde Md (8Pi51)	Central Fl	Florida West Coast	ca. 1350	Sears, 1967; Hutchinson, 1993
Tatham Md (8Ci203)	Central Fl	Florida Interior	1525–1550	Hutchinson, 1996
Weeki Wachee (8He12)	Central Fl	Florida West Coast	1525–1550	Hutchinson and Mitchem, 1996; Mitchem, 1989
Lake Jackson (8Le1)	Apalachee	Florida Panhandle	1240–1475	Jones, 1982
Snow Beach (8Wa52)	Apalachee	Florida Panhandle	ca. 1500	Magoon et al., 2001
Killearn Borrow Pit (8Le170)	Apalachee	Florida Panhandle	ca. 1500	Shapiro and McEwan, 1992
Waddells Mill Pond (8Ja65)	Apalachee/Chatot?	Florida Panhandle	ca. 1500	Gardner, 1966

Population Structure Analysis

A number of prefatory steps were taken to ensure maximum genotypic representation, as described in detail in previous publications (see Stojanowski, 2001, 2005a). These include evaluating such concerns as intra-observer error, inter-observer error, and the effects of minimal amounts of dental attrition on tooth size (age effects).[13] Data were collected for 16 odontometric variables. In the maxilla, mesiodistal and buccolingual dimensions were recorded for the first incisor, canine, first premolar, and first molar, while in the mandible, mesiodistal and buccolingual dimensions were recorded for the second incisor, canine, first premolar, and first molar. After variables with high observer error rates or demonstrable age-related attrition effects were removed, seven variables remained: maxillary premolar mesiodistal and buccolingual, maxillary molar buccolingual, mandibular premolar mesiodistal and buccolingual, and mandibular molar mesiodistal and buccolingual. Small sample sizes precluded use of anterior tooth dimensions. Focus on posterior teeth minimizes genotypic coverage; however, all tooth dimensions are generally highly correlated (Filipsson and Goldson, 1963; Garn et al., 1965b, 1968; Goose, 1963; Kieser, 1990; Moorrees and Reed, 1964; Potter et al., 1976; Schnutenhaus and Rösing, 1998) and both maxillary and mandibular and mesiodistal and buccolingual measurements were included. These different aspects of the dentition demonstrate some statistical genetic independence (Lombardi, 1975; Potter et al., 1976; Townsend and Brown, 1979) and are therefore not completely redundant.

Estimation of R-matrices, genetic distance ordinations, and phenotypic F_{ST} statistics was performed using the quantitative genetic software RMET (Relethford, 2003; Relethford et al., 1997). Heritability was estimated at .62 based on an average of previous studies from a variety of sources (see Stojanowski, 2001).[14] This is a conservative estimate that is consistent with heritabilities for craniometric (Carson, 2006; Relethford and Blangero, 1990; Sjøvold, 1984; Sparks and Jantz, 2002; Susanne, 1977), dental morphological (Scott and Turner, 1997), and anthropometric (including craniofacial) variables (Arya et al., 2002; Konigsberg and Ousley, 1995). Minimum F_{ST}s are also reported ($h^2 = 1$) because these are more comparable to those published in other studies. Where data on effective population size could be estimated, the distance ordinations are scaled to correct for the effects of differential effective population size using the method of Relethford et al. (1997).

Population Structure during the Late Precontact and Early Mission Periods

For the late precontact period, three regional aggregate samples were used: Guale, Apalachee, and Central Florida. RMET output for the late precontact period data matrix is presented in table 1.4 with the column r(ii) representing the distance of each population from the regional centroid. All effective population sizes were assumed equal for this initial analysis. The ordination of genetic distances extracted from the R-matrix is presented in figure 1.6. Precontact Central Florida and Guale populations have the lowest intersample distance (.029), followed by Guale and Apalachee (.139), and then Central Florida and Apalachee (.201). Given the time-averaged nature of these assemblages, the distance ordination should not be overinterpreted. I do note, however, that isolation by distance is not apparent. One would predict that the Apalachee and Central Florida sample are more genetically integrated based on their geographical proximity, which was not demonstrated by these analyses. The unbiased estimate of F_{ST} for the late precontact samples was .0268 (se = .0259), which does not differ significantly from 0 (p = .374).

To estimate population structure during the early mission period (1600–1650) four samples were used: San Pedro y San Pablo de Patale, San Martín

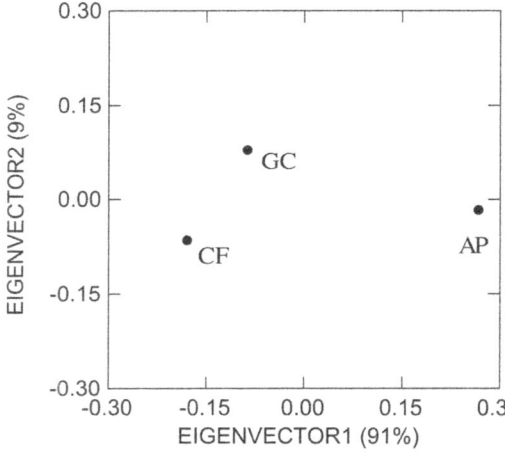

Figure 1.6. Plot of first two eigenvectors scaled by the square root of their corresponding eigenvalue for the precontact samples, nonbias corrected. CF = Central Florida, GC = Guale/Georgia coast, AP = Apalachee province.

Table 1.4. RMET output for precontact period samples

Diagonals of the R Matrix

Population	Biased r(ii)	Unbiased r(ii)	se
Apalachee	0.118608	0.018608	0.079473
Guale	0.029306	0.024760	0.008422
Central FL	0.056408	0.037177	0.024034

Note: F_{ST} = 0.068107; Unbiased F_{ST} = 0.026848; se = 0.025923.

de Timucua, Santa Catalina de Guale, and the ossuary at Santa Catalina de Santa María. Effective population sizes were estimated as follows: San Pedro y San Pablo de Patale, 15,000; San Martín de Timucua, 15,000; Santa Catalina de Guale, 1,200; ossuary at Santa Catalina de Santa María, 6,000. Results of the R-matrix analysis and intersample genetic distances are presented in table 1.5. The ordination of these distances is presented in figure 1.7. The two samples from the Georgia coast (Santa Catalina de Guale and the ossuary at Santa Catalina de Guale de Santa María) are genetic nearest neighbors; the intersample distance does not differ significantly from 0. Both Atlantic coast samples are significantly differentiated from San Martín de Timucua and mission Patale, and Patale and San Martín are likewise differentiated and roughly equidistant from one another and from the Atlantic coastal samples. Isolation-by-distance was evident during the early mission period, although somewhat weakly defined.

The unbiased estimate of F_{ST} was .0533 (se = .0175), which differs significantly from 0 (p = .027). Despite lack of statistical significance (p = .214), F_{ST} among mission provinces was double that documented during the late precontact period, increasing from .0268 to .0533. This is a noteworthy result and indicates that genetic heterogeneity increased among populations despite common inclusion with the Spanish political sphere of influence, widespread adoption of a Christian lifestyle, and documented declines in population size.

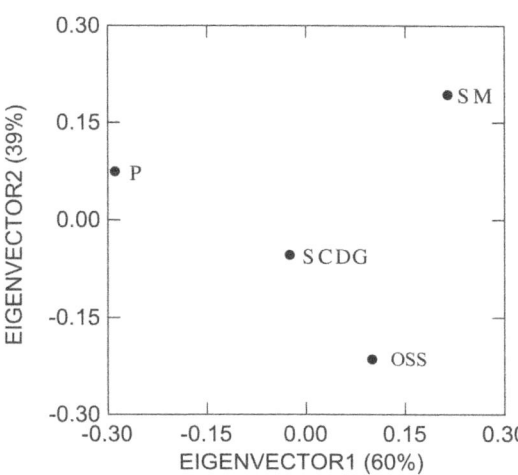

Figure 1.7. Plot of first two eigenvectors scaled by the square root of their corresponding eigenvalue for the early mission period. Genetic distances are corrected for differential effective population size. OSS = Ossuary at Santa Catalina de Guale de Santa María, P = San Pedro y San Pablo de Patale, SM = San Martín de Timucua, SCDG = Santa Catalina de Guale.

Table 1.5. RMET output for early mission period samples

Diagonals of the R Matrix

Population	Biased r(ii)	Unbiased r(ii)	se
San Martín	0.052000	0.029273	0.024891
Patale	0.079466	0.034012	0.043516
SCDG	0.125110	0.121141	0.016133
Ossuary	0.119342	0.091564	0.041688

Genetic Distance Matrix

	San Martín	Patale	SCDG	Ossuary
San Martín	—			
Patale	.175[a]	—		
SCDG	.151[a]	.187[a]	—	
Ossuary	.144[a]	.235[a]	.049	—

Note: F_{ST} = 0.082579; Unbiased F_{ST} = 0.053384; se = 0.017515.

a. Significantly different at 5% level.

Population Structure during the Late Mission Period

To estimate population structure during the late mission period (1650–1700), three samples were used: San Luis de Talimali, Santa Catalina de Guale de Santa María, and Santa María de los Yamassee. Effective population sizes were estimated as follows: San Luis de Talimali, 8,000; Santa Catalina de Guale de Santa María, 200; Santa María de los Yamassee, 1,000. Results of the R-matrix analysis are presented in table 1.6. The ordination of these distances is presented in figure 1.8. Unlike the late precontact and early mission period ordinations, there is no isolation-by-distance structure evident among late mission period populations. In fact, all three samples are nearly equidistant and there is no structure, per se, at all. Intersample distances reflect this lack of structure, which is replaced by regional genetic homogeneity (San Luis to Santa María de los Yamassee = .071, San Luis to Santa Catalina de Guale de Santa María = .013; Santa María de los Yamassee to Santa Catalina de Guale de Santa María = .037). These distances differ by an order of magnitude from those recorded during the late precontact and early mission periods among the *same* populations, and none is significantly different from 0. This pattern is particularly noteworthy considering that both Santa María Island samples were located within 40 meters of each other (Larsen, 1993; Saunders, 1993), yet the late 17th-century Guale buried at Santa Catalina de Guale de Santa María share greater affinity with San Luis, differentiated by a paltry genetic distance of .013 despite being several hundred kilometers distant from one another. Lack of population structure is consistent with estimates of F_{ST}, both suggesting that a single biological population was resident in La Florida during the latter half of the 17th century.

The unbiased estimate of F_{ST} was .0153 (se = .0056), which does not differ significantly from 0 (p = .057). This estimate is significantly lower than that observed for the early mission period (p = .039) and indicates that there was a significant reduction in between-group variability after 1650. Although not statistically significant, late mission period regional variability is roughly

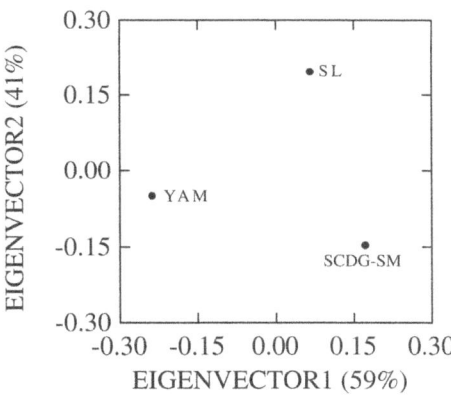

Figure 1.8. Plot of first two eigenvectors scaled by the square root of their corresponding eigenvalue for the late mission period. SL = San Luis de Talimali, SCDG-SM = Santa Catalina de Guale de Santa María, YAM = Santa María de los Yamassee.

Table 1.6. RMET output for late mission period samples

Diagonals of the R Matrix

Population	Biased r(ii)	Unbiased r(ii)	se
SCDG-SM	0.126746	0.115635	0.028084
San Luis	0.002424	0.000000	0.004342
SMdYam	0.138758	0.117925	0.040236

Note: F_{ST} = 0.019946; Unbiased F_{ST} = 0.015332; se = 0.005694.

half that documented during the late precontact period (.0153 vs. .0268). Stated in this way, F_{ST} doubled immediately after contact and then declined by over 80 percent during the late 17th century to a level half of its original baseline level.

The reduction in between-group phenotypic variability is perhaps best demonstrated when early and late mission period samples are plotted together. In figure 1.9 one can see how among-group genetic variability declined rapidly among post-1650 populations in La Florida, and importantly, this pattern is robust to the exclusion of specific samples. For example, it is not the inclusion of San Martín de Timucua or the Santa María ossuary that elevates F_{ST} during the early mission period. In fact, if you conservatively just compared well-documented Guale and Apalachee samples (Santa Cataline de Guale, Santa Catalina de Guale de Santa María, Patale, San Luis) the same pattern of temporal change in between-group variability is evident. Individuals buried at Santa Catalina de Guale de Santa María are much more similar to those buried at San Luis de Talimali than individuals buried at Santa Catalina de Guale are to individuals buried at San Pedro y San Pablo de Patale. Clearly, there was a significant change in the nature of indigenous biological interaction during the latter half of the 17th century and the results presented above are not just an artifact of sampling design, sampling bias, or sample aggregation strategies.

Nonetheless, despite the degree of morphological microevolution between early and late mission period populations, ancestor-descendant rela-

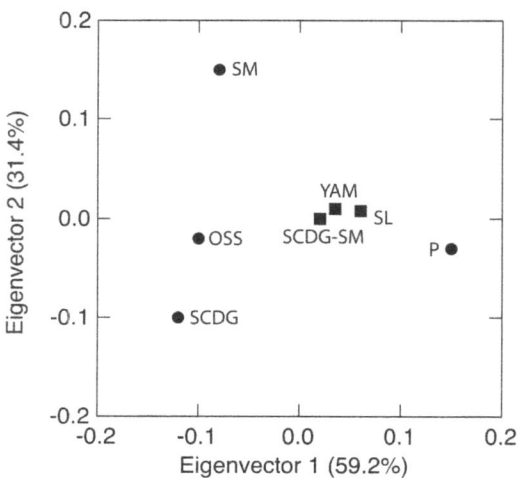

Figure 1.9. Plot of first two eigenvectors scaled by the square root of their corresponding eigenvalue for early (circles) and late (squares) mission period samples. P = San Pedro y San Pablo de Patale, SM = San Martín de Timucua, SCDG = Santa Catalina de Guale, OSS = Ossuary at Santa Catalina de Guale de Santa María, SL = San Luis de Talimali, SCDG-SM = Santa Catalina de Guale de Santa María, YAM = Santa María de los Yamassee.

tionships are apparent in figure 1.9. The late mission period genetic nearest neighbor of mission Patale is San Luis, the genetic nearest neighbor of Santa Catalina de Guale is Santa Catalina de Guale de Santa María, and the genetic nearest neighbor of San Martín de Timucua is Santa María de los Yamassee. While the last of these associations is problematic because of the unknown ethnic affinity of those people buried in the Santa María church, the data for Apalachee and Guale province samples suggest the biological changes, which are so dramatic, are not due to population replacement as has been suggested in the past (for example, Weisman, 1992: 166–68).

Interpreting Variability Statistics

As mentioned above, F_{ST} statistics have limited meaning except in a comparative context. Although the changes in regional genetic variability through time do provide critical insights into microevolutionary processes during the 17th century, an additional inference can be asked of these data: Are the mission period populations more or less genetically variable than expected given the size of the area sampled? Answering this question speaks to the relative degree of biological integration among pre- and postcontact communities living in northern Florida and Georgia which has important repercussions for interpreting these data in an evolutionary perspective. That is, by grounding the F_{ST} statistics within a broader comparative framework it is possible to differentiate the effects of genetic drift and migration on the resulting distance statistics.

To evaluate this question I recalculated phenotypic F_{ST}s for the late precontact and mission period samples using a heritability of 1 to ensure comparability with published estimates. These data are presented in table 1.7 along with comparative data pulled from the literature for which minimum phenotypic F_{ST}s were published. I restricted these comparative data to only those studies that were strictly regional in focus and based on analyses of archaeological samples. Therefore, continental or global scale analyses and those based on anthropometric data were excluded. While this reduces noise, it does not completely mitigate bias due to the different battery of traits used in each specific analysis. Therefore, the following discussion must be considered tentative.

To determine whether the mission period samples are more phenotypically variable than expected, I first normalized genetic variance estimates. To do so, I calculated correlations between F_{ST} and three variables that may

Table 1.7. Comparative minimum phenotypic F_{ST}s ($h^2 = 1$) based on cranial and dental data

Region	No. Samples	F_{ST}	Area (km²)	Time	F_{ST}/Unit	Z
Ohio Late Archaic	8	.039[a]	25920	400	.150	-.42
Ohio Late Prehistoric	7	.078[a]	96229	600	.091	-.62
Inter Miss/LW	13	.028[b]	16000	700	.175	-.35
Inter LW	7	.005[b]	16000	400	.031	-.77
Inter Miss	6	.010[b]	16000	300	.062	-.68
C. Ill Vall Miss/LW	6	.005[b]	625	700	.832	1.54
L. Ill Vall Miss/LW	7	.009[b]	700	700	1.22	2.67
L. Ill Vall LW	5	.004[b]	700	400	.557	.75
C. Ill Vall Miss	4	.002[b]	625	300	.256	-.12
Azapa Valley Coast	4	.011[c]	5000	5000	.221	-.22
Azapa Valley	3	.020[c]	5000	1500	.408	.32
Azapa Valley/Coast	7	.039[c]	10000	5000	.392	.27
Florida Precontact	3	.013	60000	200	.022	-.80
Florida Early Mission	4	.026	60000	50	.043	-.73
Florida Late Mission	3	.008	60000	50	.014	-.82

Correlations		Pearson	P-Value	Spearman
F_{ST} vs	No. Samples	.345	.207	.326
	Area (km²)	.622	.013	.607
	Time	.134	.635	.268

[a]Tatarek and Sciulli (2000).
[b]Steadman (2001).
[c]Varela and Cocilovo (2002).

affect its magnitude: the number of samples included in an analysis, the size of the geographic area represented, and the temporal duration of the samples (see table 1.7). As expected all were positively correlated with F_{ST}. As the number of samples, the size of the area sampled, and the temporal duration of samples increases so does phenotypic variability. P-values for Pearson coefficients, however, indicate that only one correlation was significant, the size of the sampled area ($p = .013$). La Florida samples were *not* included in any of the above calculations. To normalize F_{ST}s I divided each estimate by the size of the area sampled (in km²), multiplied this by 10,000 (to remove the exponentials), and standardized each estimate with a z-score transformation (subtracting the resulting mean from each estimate and dividing by the standard deviation). La Florida samples were included in these calculations. This series of adjustments results in an estimate of added F_{ST} per unit of area for each study region. Z-scores indicate the relative rank of each estimate stated in terms of standard deviation units above or below the overall mean. As indicated in table 1.7, La Florida samples, with but one exception, have the lowest z-scores of those reported, indicating the least amount of added F_{ST} per unit of area. In other words, given the size of the area sampled there is much *less* genetic diversity among populations than expected based on the comparative model. This is particularly surprising given the extent of cultural, linguistic, and political diversity included in the sampling region. Dozens of chiefdoms inhabited by individuals speaking a dozen or more languages with distinct subsistence strategies, settlement structures, mortuary practices, and material culture inventories are represented in the database. Clearly, language, culture, and biology are not isomorphic in this region. The specific meaning of this lack of diversity will be explored further in subsequent chapters, but this observed lack of diversity is telling of the extent of intertribal biological integration throughout the region.

Thus far I have identified the signals of microevolutionary change throughout the 17th century in Spanish colonial Florida. For skeletal biologists, our training dictates rigor and formality in the application of these analyses. However, from the standpoint of a bioarchaeologist, the descriptive details presented in this chapter fall short of the goal of bioarchaeological inquiry, that of contextualized historical analysis. Indeed, a critical step remains in the research process, that of interpretation. And as detailed in the following chapter, interpretation is where numeracy meets theory, or more appropriately, theories. Because the interpretive framework is critical for reconstructing the *meaning* of anthropological analyses of past populations, in the following chapter two different theoretical perspectives are considered. The first, more obvious, is evolutionary theory, while the second articulates with social anthropology and invokes the concept of ethnogenesis. Interpretations within the purview of each are offered in chapter 2 and further blended in chapter 3. The purpose is to position the study of human microevolution within the broader purview of analyses of the material past. I specifically propose that patterns of gene flow, as interpreted based on changing regional patterns of skeletal or dental phenotypic variation, reflect the complex identity discourses that occurred in the past. In other words, by studying patterns of reproductive behavior, one can reconstruct and infer the nature and extent of social interaction among communities. Increasing biological integration tracks with other forms of interaction that are socially defined. Therefore, gene flow reveals aggregate notions of affinity—how people thought about each other.

Two   *Interpretive Frameworks*

> I consider an "anthropological" approach to be one that is classically holistic as opposed to reductive; approaches questions of nature from the standpoint of historical and social perspectives in addition to natural; is sensitive to issues of gender, power, and difference instead of adopting an aloof stance; is engaged in ethical and social issues; and constantly interrogates the "given" status of natural facts and looks for ways in which they may rather be the products of human agency
> —Jonathan Marks, "What Is Molecular Anthropology? What Can It Be?" (2002: 132)

The events described in the last chapter must be interpreted within an evolutionary context while one also considers the historical trajectory of demographic and social transition in La Florida's indigenous communities. Because demographic collapse figures so prominently in the historical and anthropological literatures of the contact period, there is a distinct identifiability of cause issue that must be addressed. Genetic drift caused by population size decline is the null model such that the entire 17th century could be viewed as a period in which genetic drift was the dominant, and perhaps sole, evolutionary mechanism of significance. However, other factors must also be considered, including, for example, many more agent-based mechanisms such as migration and gene flow. Fortunately, as will be demonstrated in this chapter, the distinct pattern of changes documented among communities in La Florida belies a more complicated and nuanced explanation. A more complex yet precise account of the changes that occurred throughout the 17th century emerges. Once this evolutionary signature is defined, further consideration of the social significance of these data can then be offered. In particular, I draw from the historical ethnographic literature and find a parallel between the observed evolutionary signatures and documented stages of ethnic identity transformation in other North American colonial indigenous contexts.

Evolutionary Perspectives on Regional Genetic Variability and Patterns of Mate Exchange

In interpreting the results of the R-matrix analyses I assume that patterns of phenotypic variation represent primarily the evolutionary effects of genetic drift (population size) and gene flow (migration or mate exchange). However, given that population sizes continued to decline throughout the 17th century, genetic drift must be considered the default mechanism of evolutionary change. Genetic drift increases among-population genetic variability and decreases within-population genetic variability. The within-population expectation results from allele fixation, that is, one allele reaches 100 percent frequency and all other alleles are lost in that population. This occurs at a probability and rate that is a function of the initial allele frequency, and smaller populations require less time for allele fixation to occur. The between-population expectation results from the fact that genetic drift is a stochastic allele sampling process and, therefore, different populations are not expected to experience the effects of drift in an identical manner. In other words, different alleles become fixed in different populations that are genetically differentiated, and as a result, between-population variability increases. Determining which temporal pattern is expected is, therefore, highly dependent on knowing the boundaries and definitions of the "population," where the boundaries of the population are defined by patterns of migration and gene flow. Because we know population size was declining during the 17th century, genetic drift is expected to increase regional genetic variability if distinct populations were present in La Florida. Likewise, drift is expected to decrease regional genetic variability if a single biological population was present throughout La Florida. By holding drift constant I can directly access information on the mechanisms defining the sphere of interaction.

In chapter 1, I demonstrated that all populations in La Florida, including the late precontact samples, were not as genetically variable as expected given the size of the region and the range of cultural and linguistic diversity documented in the historical record. Genetic distance ordinations for all time periods are more reflective of long-range migration and intertribal biological integration. The transition to the mission period witnessed limited change in population structure and a doubling of genetic variability (phenotypic F_{ST}). The transition from the early to late mission period witnessed a significant change in population structure—from an isolation-by-distance structure to no structure at all—and a significant decrease in phenotypic F_{ST}. Both observations indicate that a single biological population was resident in La Florida after 1650. The most critical observation from these data is that phenotypic F_{ST} did *not* demonstrate a consistent trend through each transitional period, as expected based on changes in population size. If there was no change in migration rate through time, then decreasing population size would sequentially *increase* F_{ST} among sampled populations due to genetic drift. To the contrary, if population sizes were stable, increasing migration rates would decrease F_{ST} through time. The reversal in the overall trend circa 1650 indicates that a more complicated, multicausal evolutionary explanation is required. Drift alone does not suffice.

So what can be inferred about regional mating behavior when F_{ST} doubled from the precontact to early mission period with no change in population structure among communities that were fairly biologically integrated to begin with? This pattern could represent decreasing population size, decreasing migration rates, or most likely both. That migration rates also declined in concert with decreasing population size is supported by four observations. First, the limited genetic diversity among populations and the isolation-by-distance population structure indicate that a broad regionally defined system of migration and mate exchange was in place from the late precontact through late mission periods. Second, genetic distances and estimates of between-population genetic variability were corrected for the expected effects of genetic drift by scaling the R-matrix by effective population size. This adjustment should theoretically reconstruct a population structure due solely to migration. Third, changes in phenotypic variances within provinces across the contact-period transition do not uniformly demonstrate a reduction in within-group variability (Stojanowski, 2001, 2005a), as predicted by genetic drift. In fact, Stojanowski (2001, 2005a) documented an increase in genetic diversity from the precontact through early mission period in Guale province, and no significant change in genetic diversity across the contact transition in Apalachee province. Therefore, the effects of drift may not be as severe as one might expect based on population size enumerations. And finally, there are substantial historical data to suggest that migration rates *were* affected by the missionization process. Specifically, I propose that changes in settlement structure and the sociopolitical zeitgeist obviated existing late precontact and protohistoric period cultural interaction patterns causing much of the value system to lose saliency after missionization. This perspective is further developed in chapter 4.

The transition from the early to late mission period requires a far simpler explanation. The dramatic decline in regional genetic variability as well as the loss of an isolation-by-distance population structure indicates that a relatively homogenous biological population was present throughout the mission communities of La Florida. Although a decline in genetic diversity is expected due to population size decline and genetic drift, it is only in the context of an expansive and cohesive migration and mate exchange network that genetic drift would *decrease* regional genetic variability. Because the preceding temporal phase witnessed the opposite pattern (genetic variability increased among Apalachee, Guale, and Timucua populations), the early mission period can be characterized as one of genetic microdifferentiation in the context of declining local population sizes or migration rates. The reversal in this trend, however, points to a change in the structure of migration after about 1650. Gene flow was more expansive.

A Model for Colonial Ethnogenetic Transformation

Maintaining a strict interest in evolutionary research would curtail this discussion at this point. The data have been analyzed and interpreted with respect to the predictions of evolutionary theory, and specific mechanisms have been identified which explain the patterns in the data. However, gene flow in humans is not quite the same thing that it is in other mammals. Gene flow elicits a more nuanced consideration of the underlying subtext of what these changing patterns of variation meant to those who participated in the process. Humans are not irrational reproducers driven solely by the selective drive to increase fitness; and mate exchange, therefore, has a very powerful social component in its patterning. In this sense, the mid-17th century was a turning point for the populations of La Florida and something very different was happening, which I argue is not just a biological or evolutionary phenomenon. In fact, I propose that these changes in biological patterns are reflective of concurrent changes in the social fabric of indigenous societies along the frontiers of European expansion. New identities were emerging in the wake of demographic collapse, and ethnogenesis was occurring as new notions of ethnic affinity were defined and inculcated by previously distinct tribal communities. More specifically, I argue that this was a fusion-based form of ethnogenesis in which disparate and distinct ethnic communities were merging into a new and unique social identity—that is, an ethnic amalgamation, sensu Horowitz (1975), was occurring.

At first glance, this association between biological variation and social variation may seem facile. And, in fairness, I am unaware of previous research that explicitly links a microtemporal evolutionary process to the concept of ethnogenesis. However, invoking a social theoretical interpretation is consistent with Nancy Hickerson's (1996) generalized model of fusion-based ethnogenetic change, which she likened to "life-cycle transitions" (Hickerson, 1996: 70). The initial phase, *separation*, involves "the negation or severing of their existing group loyalties"; during the *liminal* phase, "surviving—usually dysfunctional—social and/or economic ties wither away, and alternative connections are initiated"; and *reintegration* occurs when a "new identity is consolidated, affirmed through ritual and the adoption of a validating mythology" (Hickerson, 1996: 70). This is a familiar transformational model (Turner, 1969; Van Gennep, 1960) and is one that encompasses the more specific heuristic schemes of Sharrock (1974) and Albers (1996)(figure 2.1), both of which represent different levels of ethnic incorporation that fall within the liminal category of Hickerson (1996). In other words, after separation but before reintegration there are several levels of interdependency that distinct social aggregates can assume. It is during this liminal phase that communities are "creating a sense of cohesion" and at the same time engendering "external differentiation, generating a belief in difference from others" (Bell, 2005: 447).

These models are heuristic devices and represent degrees of ethnic incorporation rather than an evolutionary framework comprising stages in the ethnogenesis process (see also Eriksen, 2002: 43–44; Handelman, 1977; McKay, 1982). Hickerson's is the most general; however, Sharrock and Albers both present a fusion-based model of ethnic emergence in which communities pass through a period of mutual alliance and shared interests, followed by coresidency and intermarriage (codification of these shared interests), with the final phase representing ideological and nominal reification of the new group identity. Distinct ethnic groups form the base of the model and a single, aggregate ethnic group forms the top. However, it is important to stress that there is no temporal component, that is, a "push" mechanism, which moves ethnic groups toward complete assimilation. Nor is there nec-

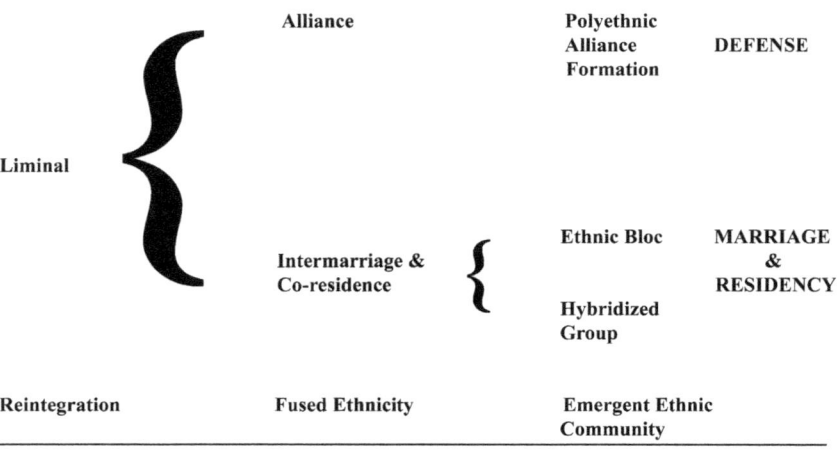

Figure 2.1. Ethnogenetic models derived from the colonial ethnographic literature.

essarily a stepwise unilineal process being presented. In fact, communities may oscillate along this spectrum of incorporation, moving up (merger) or down (fission) the scale as circumstance permits. These social identities may always be in a state of "becoming" such that the postliminal stage is never achieved. In fact, to assume such a temporal dimension imposes a decidedly Western cosmogony on human social behavior.

In table 2.1 details of Hickerson's generalized model are linked with biological signatures for each phase of the ethnogenetic life cycle. During the first phase, *separation*, existing social identities lose saliency because of changes in the social-political structure of relationships within and between ethnic groups (Hickerson, 1996). As a result, new social connections are sought, often in an experimental manner, as individuals and communities "test the waters" situationally (Barth, 1969; Cohen, 1969, 1974; Cohen, 1978; Glazer and Moynihan, 1975; Nagata, 1974). Here it is proposed, in recognition of similar assimilation studies using material culture (e.g., Bell, 2005; Hardesty, 1999; McGuire, 1982, 1983; Mullins and Paynter, 2000; Praetzellis et al., 1987; Staski, 1987; Voss, 2005), themselves based on classic debates about the appropriate interpretation of style in the archaeological record (Conkey,

1990; Hegmon, 1992; Hodder, 1982; Sackett, 1982, 1985; Wiessner, 1983, 1984, 1985, 1990; Wobst, 1977), that the separation phase of ethnogenesis and its associated disruption of social organizational mechanisms of intercommunity integration results in a decline in regional genetic integration. This was demonstrated by the analysis of early mission period samples presented in chapter 1. Genetic diversity among these populations nearly doubled in comparison to late precontact populations sampled from the same geographical regions, but without a significant change in population structure.

Table 2.1. Ethnogenetic model of biosocial transformation

Phase	Social Correlates	Biological Correlates
Separation	Existing integrative mechanisms disappear	Increased between group heterogeneity
Liminal	New social connections forged	Decreased between group heterogeneity
Reintegration	Validation of identity through shared ideology	No predicted biological changes

As the liminal period commences and new social connections are established, I propose, in analogous fashion with historical archaeological approaches (e.g., Bell, 2005; Voss, 2005) and the predictions based on recorded observations of Albers (1996) and Sharrock (1974), that these new social connections will also have biological correlates. Patterns of regional genetic variability will indicate increasing homogeneity, reflective of new patterns of intercommunity integration established in response to external stimuli that initiated the ethnogenesis process. As demonstrated in chapter 1, late mission period populations experienced an 80 percent *decline* in genetic diversity and demonstrated a complete lack of biological structure among missionized populations. This is most parsimoniously interpreted as evidence of extensive and widespread mate exchange among mission populations throughout La Florida resulting in a complete redefinition of population boundaries and thus allowing the effects of drift to manifest as a within-population signal (decreasing diversity).

The third phase of ethnogenesis, *reintegration*, has no hypothesized biological analogue, except perhaps a further decline in genetic diversity, but is instead based on reification of group ethnonyms, a self-ascribed consciousness, and a belief system constructing a common origin and engendering feelings of a common historical fate. I specifically propose that this phase had not been completed during the 17th century and was, in fact, preempted by the English assault on the Florida colony (see chapter 1). This resulting disruption of endemic social adaptive processes prevented reintegration—it prevented development of a self-referential ethnonym that is required for recognition in the European historical annals. Rather, the English raids and resulting diaspora of Spanish Catholic Indian communities almost completely obscured evidence that these processes were ongoing, a statement in and of itself subject to additional scrutiny in the chapters that follow.

At this point I have documented a pattern of microevolutionary change, argued that the pattern primarily reflects changes in mate exchange practices through time, and linked the specific trajectory of integrative mechanisms to an analogous model from the historical ethnographic literature on postcontact Native American ethnogenesis. However, before additional evidence that ethnogenesis was ongoing among communities in Florida and Georgia is fully considered, the divide between the "biological" and "social" realms of human experience must be explicitly linked. In particular, it is critical that biodistance data be embedded within existing approaches to the material remains of the past and articulated with current conceptualizations of human social identities, and ethnic identity in particular. I turn to this social theoretical literature in the next chapter.

Three

Ethnogenesis, Social Identity, and Human Biology

A Bridging Model

> Proponents of ethnogenetic models of human history and evolution argue that human societies periodically reorganize themselves and that the resulting new social formations are likely to have their "roots" or "origins" in several antecedent societies. . . . The resulting patterns of diversity in biology, language, and culture can be said to be more like a "tapestry" than a "family tree."
> —John Edward Terrell, *Archaeology, Language and History* (2001: 31)

Identity is a topic of considerable importance in anthropology. Its study is one of the unifying research themes within the social sciences because multiple data sources (historical, ethnographic, archaeological, and less prominently, biological) can be brought to bear about a subject with saliency in the modern world. Identity can be tribal, regional, national, or religious in practice, in addition to gender-, status-, class-, or age-based, but it often refers to ethnic groups, their boundaries, composition, and relationships with other forms of social order. And this, the concept of social order, is at the heart of identities research. Beyond ascribed kinship relationships, the corpus of personal identities in large, plural societies allows individuals to relate to one another. We have identities, not an identity. Just as we assume different social personas, with concomitant changes in role, throughout the course of any given day, we also assume different identities as social and economic stimuli change. These forms of relational organization are, therefore, highly adaptive in the complex social world of humans. They broaden the basis of belonging and represent, in some way, "modern clans" in which historical connections are uncertain but shared nonetheless among members of society. Our human identities simultaneously divide us into communities while building solidarity within them. To Arnold Epstein, social identity is "the process by which the person seeks to integrate his various statuses and roles, as well as his diverse experiences, into a coherent image of self" (Epstein, 1978: 101), while to Barbara Voss, social identity refers to the "many ways that individuals and groups of people are taxonomically categorized through socially constructed relationships of difference" (Voss, 2005: 461).

Ethnic identity is considered by some the dominant identity of an individual, superseding sentiments based on class, age, or gender (Barth, 1969; Cohen, 1978; Eriksen, 2002; Horowitz, 1975; Roosens, 1989), and ethnic groups are defined on the basis of sharing a perceived or real common origin, being self-referential, and being distinct from similar groups of equivalent function (Hicks, 1977: 3). In addition, ethnic identity is "segmentary" (Cohen, 1978: 387); multiple inclusive levels exist at different scales of social integration and therefore it is an abstract organizational mechanism (see Handelman, 1977; Jenkins, 1997; McKay, 1982; Roosens, 1989: 19–20) even within a single structure of relationships. Since Barth's (1969) influential paper on ethnic boundaries, anthropological research on ethnicity and ethnic groups has been concerted. However, as noted by Albers (1996) and Shennan (1989), scholars have tended to focus on definitional, functional, and synchronic issues of ethnicity rather than ethnic genesis, how groups change their identity in reference to external forces, or how multiethnic societies originate. Therefore, while ethnicity remains a popular concept in anthropology (see Banks, 1996; Cohen, 1978; Eriksen, 2002; Jones, 2002), and, indeed, has become a raison d'être for a significant component of social anthropology, ethnogenesis has received less attention.[1]

Etymologically, ethnogenesis is the process through which ethnic identities emerge or evolve.[2] However, as defined by William Sturtevant, ethnogenesis is simply "the establishment of group distinctness" (Sturtevant, 1971: 92), however that group is defined.[3] In other words, ethnogenesis is not just about ethnicity but any form of communal ascription at various scales of integration (see in particular Jenkins, 1997, but also Canuto and Yeager, 2000; Cohen, 1985; Isbell, 2000). As such, whether a social identity is community or regionally defined, tribal, nationalist, based on shared ideology, or imbued with an "ethnic" quality may, in some ways, be irrelevant. These forms of identity likely experience similar rules of development, differing perhaps only in degree of permanence or a mythological charter of common ancestry.[4] This position is not meant to mitigate the important definitional and historical differences between ethnic groups and other forms of social organization (see discussion in Jones, 2002: 40–55). However, the interest here is not in determining whether indigenous communities in La Florida represented a specific kind of ethnic group or degree of ethnic incorporation (sensu Albers, 1996; Handelman, 1977; McKay, 1982; Sharrock, 1974), but rather to infer that the process of ethnic fusion was commencing, that a shared group identity was emerging.

Focus in this volume on the ethnogenetic process also divorces my interests from the broader literature on ethnic identity that concentrates on understanding how ethnic groups function within larger sociopolitical contexts; why ethnic groups may have initially formed in human history; the mechanisms responsible for burgeoning ethnic consciousness among modern migrant communities; and the relationship between ethnicity and nationalism, class, race, and prejudice. Therefore, I make no pretense of contributing novel theoretical perspectives on *why* ethnic groups form (I do not solve the great enigma of "ethnicity"), so it is somewhat pedantic and unnecessary to rehash earlier debates. One of these debates deals with whether ethnic sentiments are primordial (Bromley, 1974; Connor, 1978; De Vos, 1975; Epstein, 1978; Geertz, 1963; Isaacs, 1974; Keyes, 1981) or instrumental (Barth, 1969; A. Cohen, 1969, 1974; R. Cohen, 1978; Glazer and Moynihan, 1975; Nagata, 1974),[5] a theoretical dichotomy succinctly encapsulated as "communities of culture" (Cornell, 1988: 20) that are sentimental and based on an "intuitive bond" (Espiritu, 1992: 4) engendering rootedness, versus "communities of interests" (Cornell, 1988: 20) that are situational and strategic and rooted in collective action for material gain. Another is Bentley's (1987) consensus-building practice theory, which views ethnic identity as resulting from a "subliminal awareness of objective commonalities in practice" (Bentley, 1987: 27; see also Jones, 2002, 2007; McKay, 1982; Orser, 2004), itself based on Pierre Bourdieu's (1977) concept of *habitus*. Still others deal with whether ethnic identity is objectively or subjectively ascribed (it is, or can be, both—Barth, 1969; Handelman, 1977; Naroll, 1964; Cohen, 1978), or the relationship between individual identity and collective ethnicity (Banks, 1996; Bentley, 1987; Eriksen, 2002; Jones, 2002). Numerous publications that summarize these different perspectives do a much more thorough job than I could ever offer. Where appropriate, details from these theoretical perspectives are incorporated into the chapters that follow; however, no primer is offered herein. Although, admittedly, one cannot completely understand genesis in the absence of function, concentration on function alone assumes that ethnic groups, ethnicity, or ethnic consciousness is a "monothetic . . . class of phenomena" (Comaroff, 1987: 302). This is simply not the case, and there can be no single path of creation when the end product (ethnic identity) is not itself uniform structurally and functionally, or when perpetuating factors (those that sustain an identity) could be distinct from originating factors (those that caused the initial emergence). That is, the economic or political factors that cause ethnic boundaries to shift or to manifest are quite different from the cultural and ideological processes required to maintain them.

The complexity of formative and sustaining factors is matched by the historically contingent nature of many cases of documented ethnic emergence. Why ethnogenesis occurs in any particular situation is highly dependent on very specific historical circumstances with contingency upon contingency interacting in a constant process of transformation. The acts of individuals and the provisions of specific treaties can be hugely significant on long-term historical trajectories, and the complexity and specificity of factors contributing to ethnic transformation inhibit development of broad generalized models. As noted by Eugeen Roosens, "there is no single, uniform process of ethnogenesis. . . . The variety of processes and meanings . . . constitute, in their own right, an integral part of a complex reality. No deep understanding or explanation is possible *without a painstaking reconstitution, case by case*" ([emphasis added] Roosens, 1989: 149). Indeed, ethnogenesis can result from either the fission or fusion of ethnic elements, with distinct repercussions associated with the permutations of each (see table 3.1).

Table 3.1. Models of ethnogenetic change

Division	A >>> B and C	2 new groups form
Proliferation	A >>> A and B	fission of 1 group
Amalgamation	A + B >>> C	2 groups fuse
Incorporation	A + B >>> A	minority fuses into majority
Panethnogenesis		higher order identity created
Transformation	A >>> B	significant change in group identity

Source: Horowitz (1975); Arutiunov (1994).

Historical ethnographic analyses of ethnogenesis support this sentiment. For example, for previously enslaved African maroon communities, social upheaval and the harsh abuses of plantation life fostered the development of new ethnic sentiments in polyethnic communities whose initial raison d'être was simply survival (Bilby, 1996; Cordell and Yannie, 1991; Kopytoff, 1976). For many Native American groups, such as the Jumano, Kiowa, and Cheyenne (Hickerson, 1996; Moore, 1994a, b, 2001), ethnogenesis entailed processes of accommodation, adaptation, merger, and/or fission as colonial policy and epidemic disease effected a new sociopolitical order which resulted in a "Hobbesian war of all against all" (Hill, 1996b: 5). Sharrock (1974) highlighted the impact of intergroup warfare, often related to changes in local economies brought about by European traders, in her analysis of Cree and Assiniboin ethnic incorporation based initially on shared horsemanship and a long-standing alliance that pitted them against other more powerful indigenous groups such as the Sioux. For the Houma (Davis, 2001) and "neo-Chumash" (Haley and Wilcoxon, 2005), ethnogenesis and the adoption of an "Indian" identity was directly related to legal and cultural attitudes toward African Americans and the active negotiation to be classified as "non-Black" in a world in which they could also not be "White." In Africa, ethnogenesis is intimately linked with indirect rule and multiethnic nation-state building (Levine and Campbell, 1972; Vail, 1989; see also Keller, 1995) in which tribalization strengthened preexisting social groups to prevent pan-African unity movements from forming (Mahoney, 2003).[6] All of these examples, few among many, demonstrate a commonality in their emphasis on understanding the historical and social context in which identity discourses occur. They also share a concern with "composition" and "construction," that is, how human social groups are organized, interrelated, and ultimately defined in a historical sense.

Theoretical Basis of the Biological Ethnogenetic Model

Different from understanding questions of how ethnic sentiments develop (practice theory, objective/subjective criteria) or why they may evolve or disappear (instrumentalism) is defining how one *recognizes* that ethnogenesis is occurring, an operational approach many avoid in favor of synchronic, functional considerations. Here, I propose that evidence of ethnic emergence manifests, as in other materially and socially visible markers, in patterns of gene flow that reflect human behavior and action at the community level and therefore serve as proxies for communal recognition of "self," "us," and "other." In other words, gene flow, intermarriage, or mate exchange between populations reflects changing emic (subjective) definitions of the people themselves; genetic and phenotypic signatures *passively* signal ethnic emergence as part of the development of a communal "esprit du corps." Increasing exclusivity in patterns of mate exchange *may* suggest ethnic factionalism and devolution into distinct identities, whereas homogenization of genetic exchange networks *may* reflect a broadening basis of identity which may ultimately lead to ethnogenesis in a reticulate manner. The former suggests ethnogenesis through fissioning, where new ethnic groups emerge due to divergent interests (for example, the Seminole from the Lower Creeks—Sturtevant, 1971), whereas the latter reflects ethnogenesis through the fusion of previously distinct ethnic groups (for example, many postcolonial Native American tribes—Hill, 1996a, and contributors therein).

Two critical details should be emphasized here. First, it is important to stress the temporal component of these expectations. Indeed, I propose that interpreting patterns of phenotypic variation in a temporally static framework may tell us very little about the ethnogenetic process or ethnic group composition. Only by reconstructing how patterns of mate exchange changed through time can we infer similar changes in ethnic or community sentiments, a perspective which accords well with classic ethnographic observations on the permeability of ethnic boundaries and the situational nature of ethnic groups which may have little actual content in specific social settings. Second, I am not in any way proposing that the biological signatures of ethnogenesis are deterministic. That is, I very carefully defined this model as one of potentiality rather than strict causation. While I return to this issue below, I think it is important to highlight this point repeatedly throughout this book. And, in fact, the bulk of this book attempts to provide additional contextualization of the biological changes documented in chapter 1, which supports the interpretation of ethnogenetic transformation proposed herein.

Ethnogenesis Does Not Equal Origin

This microevolutionary approach departs from past biological or evolutionary consideration of ethnic identity as well as past consideration of the biological history of social identity groupings. This literature, produced both by skeletal biologists and molecular anthropologists (Derenko and Shields, 1998; Hill et al., 2006; Hurles et al., 2005; McEvoy et al., 2004; Posukh et al., 1998; Ricaut et al., 2006; Shubin et al., 1997; Udina and Rautian, 1994; Udina et al., 1985; Yao et al., 2000), often refers to population origins, but not in geographical terms (such as peopling of the New World or peopling of Polynesia), but in terms of the biological composition of an often nebulously defined contemporary social aggregate (e.g., Barbujani and Sokal, 1990; Batista et al., 1995; Cavalli-Sforza, 1997; Cavalli-Sforza et al., 1988, 1992; Hill et al., 2006; Hurles et al., 2005; Hunley and Long, 2005; Hunley et al., 2007; Ioviță and Schurr, 2004; Kolman and Bermingham, 1997; Kolman et al., 1995; North et al., 2000). Phenotypic approaches are historically precedent but generally use research designs similar to those of anthropological geneticists, considering, for example, the ethnogenesis of Icelanders (Pálsson, 1978), Thracians (Boev, 1980), Central Asian tribal groups (Oshanin, 1964), and dozens of Siberian ethnic groups (Michael, 1962). More recent bioarchaeological contributions have moved beyond typological methods, although concern for migration, affinity, and contributing population composition remains (e.g., Hallgrímsson et al., 2004; Hemphill and Mallory, 2004; Moiseyev, 2001; Ossenberg et al., 2006; Zakrzewski, 2007).

In every case, discussion focuses on population composition and the contributing origin of represented types. As such, these studies are self-contained and do not directly contribute to or draw from social theory. For this reason a different biologically based ethnogenetic model must be developed.

The Body as Material Culture

The approach used in this book draws its operational framework from historical archaeology and the analysis of stylistic variation in the past. Although historical archaeologists have long considered culture contact, acculturation, and ethnic identity, explicit inferences of ethnogenesis have been more limited but nonetheless highly influential on this research model (Hardesty, 1999; McGuire, 1982, 1983; Mullins and Paynter, 2000; Praetzellis et al., 1987; Staski, 1987; see also Upton, 1996). Two recent case studies are exemplary of this approach, both powerful yet simple in application.

Barbara Voss (2005) documented the emergence of a unified "*Californio*" identity at El Presidio de San Francisco based on analysis of material culture, architectural structures, and dietary preferences. Despite the biologically diverse population in residence at the presidio, inhabitants demonstrated an increasingly similar material culture repertoire (plain ware ceramics, similarity in cooking implements and techniques, and similar architectural styles) and actively chose diets focusing on gruels lacking specific ethnic culinary symbols. Based on temporal changes in these markers indicating increasing homogeneity, Voss (2005) inferred ethnogenesis was occurring as the communal self-referent *gente de razón* replaced the divisive "mixed-race" structure of the *sistema de castas*. Over the course of three decades, the residents of this military establishment suppressed the objectifying *castas* categorization while simultaneously eschewing material symbols of local indigenous populations, both signaling, via emblemic elements, their cohesion and solidarity. Trends in material culture signaling increasing homogeneity were directly related to the zeitgeist of burgeoning *Californio* community consciousness embedded within local and regional historical processes.

Alison Bell (2005) similarly interpreted patterning in architectural features as evidence of "White" ethnogenesis among 17th-century residents of the Chesapeake Bay region. In this case, the removal of slave laborers from the household, the use of hall-parlor-style dwellings, homogeneity in house size and style (i.e., lack of conspicuous consumption), and earth-fast construction techniques symbolized unity among European colonials via their shared experiences, social expectations, and common historical fates; they also engendered community solidarity by instituting a system of interconnectivity and symbiotic cooperation. Earth-fast construction techniques require considerable maintenance, and Bell proposes this was an intentional decision on the part of European residents of the Chesapeake to foster economic codependency.

The case studies of both Voss and Bell highlight the importance of practice in building community identities, linking microscale practices to macroscale processes (see also Lightfoot et al., 1998; Pauketat, 2001, 2003; Shennan, 1993). In addition, although neither Bell nor Voss pays homage to classic debates on the symbolic components of material culture and style (Conkey, 1990; Hegmon, 1992, 1998; Hodder, 1982; Sackett, 1982, 1985; Wiessner, 1983, 1984, 1985, 1990; Wobst, 1977—see also Jones, 2002; Shennan, 1989), it is clear that elements of these perspectives on the relationship between social identity and material culture are being applied. However the papers by Bell and Voss differ from the corpus of research on material culture style as symbolic of social identity because of the focus on diachronic process, which is quite distinct from synchronic consideration of spatial distributions of material culture and their presumed social meaning (see Hegmon, 1992; Hodder, 1979; Jones, 2002; Shennan, 1989).

In a similar vein, I am not interested in identifying the boundaries of ethnic groups in Spanish colonial Florida (an exercise in allocation) but in identifying the signatures of ethnogenesis. The former (ethnic groups) are abstract, malleable social categories beyond simple description; the latter is the process by which these entities come into existence and transform through time. Sîan Jones clarifies the distinction between synchronic and diachronic approaches as one of scale and timeframe: "within a diachronic contextual framework it may be possible to pick up the *transformation* of habitual material variation into active self-conscious ethnic symbolism . . . on the basis of changes in the nature and distribution of the styles involved" (emphasis added, S. Jones, 2002: 126). This is exactly the model adopted by colonial researchers interested in ethnogenesis (Bell, 2005; McGuire, 1982, 1983; Voss, 2005) and which I co-opt in a literal interpretation of the "body as material culture" perspective (Sofaer, 2006). However, this analogy should not be drawn too thinly or literally. Style is not monolithic and unidimensional, and the production of material culture variation is clearly governed by rules different from those determining phenotypic variation. Although evolutionary archaeologists may disagree with this statement, the approach adopted in this book does not assume that phenotypic and technological data experience the same rules of development. The goal is simply to normalize the study of human skeletal remains within the broader realm of research on residues of the past.

In this sense, phenotypic variation may be completely passive and not symbolic in any active sense of the word because individuals have no knowledge of the material expression of these underlying social identities. That is, the production of style with identity symbolism may be conscious or unconscious, active or passive, but it is usually visible to the actor and the observer. Phenotypic variation, tooth size in particular, is not visible to the actor or the observer and cannot be viewed as actively messaging within an identity discourse framework (as opposed to cranial and dental modification). Nonetheless, phenotypic variation *is* reflective of the choices one makes within social constraints in the field of reproduction. It is the product of a specific "production sequence" (*chaîne opératoire*) that is laden with identity symbolism. One's upbringing, heritage, education, early life experiences (the production sequence)—all of these determine or inform in some way the decisions one makes about reproduction, or in the Western sense, marriage. As such, it is easy to envision the complexity of social meaning imbued within something so simple as changing patterns of genetic variation within an extended breeding network.

Tensions between Biology and Culture

At this point it is important to readdress the topic of intentionality and determinism. The temporal homogenization of material culture patterns, practices, and architectural styles is read as a "signal" but not necessarily an overt cause of ethnogenesis. Of course, when atomized into discrete analytical units this nuance is clear enough; however, more inclusive consideration of "practice" as the process of social production of community identity most certainly considers the mundane, everyday, habitual, overlooked, and automatic behaviors as wholly part of the ethnogenesis process (Bentley, 1987; Lightfoot et al., 1998; Orser, 2004; Pauketat, 2001). Nevertheless when biological data are considered, the distinction between active causality and passive representation is less apparent, and there is an enduring legacy of controversy surrounding the interface of evolutionary biology and social phenomena such as ethnic groups. Therefore, stating my position clearly is essential.

Despite ubiquitous anecdotal mention in historical ethnographic studies that intermarriage is a component of ethnic emergence (see examples in Hill, 1996a, and also in most case studies of ethnogenesis based on ethnographic and ethnohistoric data)—and in fact, both Albers (1996) and Sharrock (1974) codify the importance of intermarriage in their semievolutionary ethnogenetic models for North American populations (see chapter 2), the role of mate exchange is infrequently explored in an explicit microevolutionary framework. However, the wholesale appropriation of a social constructionist perspective within anthropological thought has completely jettisoned biology from discourse on social identities. In fact, discussion of genetics or biology in the context of ethnicity will often provoke a negative response under the assumption that biological determinism, naive primitive isolationism, or sociobiological sentiments are being promoted (see van den Berghe, 1986, for a rebuttal of these criticisms). More frequent, of course, is the conflation of ethnicity with race, or as summarized above, the somewhat facile equating of social categorizations with biological populations. As noted by Chapman (1993a: 1), "social and biological approaches are often pursued independently of one another, are often mutually ignorant, and are sometimes mutually hostile." Marcus Banks, commenting specifically about the sociobiological perspective of van den Berghe (1978, 1981, 1986), feels that biological approaches to ethnicity "constitute a rather self-contained subfield" (Banks, 1996: 2), while Richard Jenkins (1997: 48) demurred that examination "of the relationship between biology and culture, has receded from the anthropological agenda." By eschewing typological determinism and positioning the analysis of phenotypic variation within the broader scope of material culture studies, it is my hope that this research moves beyond this suggested impasse.

To do so, sociobiological and biosocial interpretive frameworks must be explicitly differentiated (Chapman, 1993a, b). The former views ethnic groups as an extension of kin groups in which nepotistic behavior is favored over the long term by natural selection, and ethnic sentiment therefore fosters inclusive fitness (van den Berghe, 1978, 1981, 1986). In this sense, genetic homogenization is seen as a requisite precondition for ethnogenesis, which has the unfortunate effect of conflating "social community" with "biological population" as well as adding an unnecessary dimension of determinism and intentionality to marriage and mate exchange. There are many criticisms of this primordial perspective. It is biologically deterministic, it provides simplistic solutions to complex issues, it facilely parses society and culture, and there is a seemingly insurmountable implication that humans are programmed for conflict and biological discrimination (see Jones, 2002: 71–72; McKay, 1982; Reynolds, 1980; Rubin, 2000; van den Berghe, 1986).

Biosocial approaches, on the other hand, view the biological and social realms as synergistic and divorce human behavior from the proprietary grasp of Darwinian selection. Although both sociobiological and biosocial approaches are based in evolutionary theory, the latter does not assume an adaptationist perspective nor does it embrace intentionality. The focus is on the social symbolism of gene flow, not natural selection. The underlying basis of material patterns of variation is social process, not subconscious psychological motivation and kin selection. Gene flow is not relegated to a mechanical demographic parameter (migration) but is instead used as a starting point to try to understand the social underpinnings which effected a change in migration and marriage practices.

In this sense, while this book does not contribute a novel theoretical perspective or summarization of the ethnicity concept, one of the broader goals is to correct the perceived disjuncture and ameliorate the tension between biological and social aspects of studies of ethnic identities. In particular, the biosocial model presented here is not based on racial models, patterns of genetic ancestry and descent, population origins, sociobiology and kin selection, or any of the other divisive and controversial uses of biology in the social realm. Ethnicity is not a neologism and euphemism for race.

Active or Passive Contributions to Ethnogenesis

The critical distinction is whether gene flow represents an active or passive part of the ethnogenesis process. For example, the strongest active position proposes that intermarriage and mate exchange *cause* ethnogenesis through the pursuit of common material interests during the course of kin selection. This is the sociobiological perspective espoused by van den Berghe (1986). That intensively competitive ethnic groups are sometimes genetically similar—for example, Israelis and Palestinians in the Mid-East (Arnaiz-Villena et al., 2001) or the Oromo and Amhara in Ethiopia (Tartaglia et al., 1996)—should give sociobiological primordialists some pause.

It is, however, possible to retain active causality in the bridging argument and not assume a sociobiological perspective. For example, the conflation of social identity categories in the context of expansive systems of mate exchange and intermarriage may obviate existing classification schemes and add considerable complexity to social rules and norms of identification. Social cohesion may manifest through the bonds of marriage (symbolic active) or the difficulty of maintaining burgeoning complexity of ethnic labels when new generations of "mixed" ethnic offspring are born (passive). In either case, gene flow between populations contributes to the ethnogenetic process but does so in a completely different manner from that of the sociobiological model. It lacks the sociobiological elements of motivation and intent and may be more palatable to social scientists as such.

An alternative perspective uses a passive bridging argument and is much more in line with the previous discussion of material culture and style. In this perspective gene flow (genetic homogenization) between previously distinct populations could merely *reflect* changing sentiments and dispositions among those involved in the social field of interaction but contribute no causality to the process itself. As such, gene flow is not a cause of ethnogenesis but a result of it. It is the behavior itself, the process of mate choice, that signals a change in emic sentiment redefining "us" but not necessarily "them." This approach rejects any a priori notion that biological homogenization necessarily *causes* feelings of shared ethnic solidarity among previously distinct communities and simultaneously eschews notions that ethnic groups are, by definition, genetically homogenous (read, biological population) or that genetic homogenization is a *requisite* condition for identity transformation. As such, the passive model is consistent with the corpus of social theoretical constructions of ethnic identity and actually serves well the ubiquitous notion that ethnicity is defined in some way by a belief in a common ancestral origin. Indeed, the conflation of "ethnic group" with "biological population" is contrary to the observed permeability of ethnic boundaries and also assumes stability in population composition, neither of which is consistent with the situational nature of ethnic identities. Because ethnic groups can be genetically homogenous or heterogeneous at any particular point in time, only diachronic consideration of changes in panmixia are salient. Only in a comparative, temporal perspective may changes in genetic variability indicate similar changes in shared communal sentiments.

In this chapter I have attempted to relate patterns of gene flow to feelings of ethnic solidarity. As this approach to biodistance analysis in past population is unorthodox, the interpretations set forth would be considerably strengthened if subsidiary, independent data corroborated this interpretation of 17th-century community organization in Spanish colonial Florida. Specifically, is there archaeological or historical evidence that better contextualizes the biological analyses? Is there support for the inference that the first half of the 17th century was one of ethnic demobilization, associated with a change in community integrative mechanisms which had previously defined interpolity interactions? Likewise, is there independent support for the interpretation that the second half of the 17th century witnessed the initiation of new and exploratory intercommunity interactions which united the Spanish mission populations into a monolithic social community with a burgeoning shared identity? The next three chapters address these questions in greater detail, focusing at first on the separation phase between 1600 and 1650 and then proceeding to consider the liminal phase in chapters 5 and 6.

Four

From Tribe to Ethnic Group, or from Ethnic Group to Tribe

Ethnic Demobilization during the Early Seventeenth Century

> To speak of an ethnic group in total isolation is as absurd as to speak of the sound from one hand clapping. By definition, ethnic groups remain more or less discrete, but they are aware of—and in contact with—members of other ethnic groups. Moreover, these groups or categories are in a sense *created* through that very contact. Group identities must always be defined in relation to that which they are not.
>
> —Thomas Hylland Eriksen, *Ethnicity and Nationalism* (2002: 10)

That ethnogenesis was occurring in Spanish colonial Florida is not a novel observation. Nonetheless, indigenous ethnogenesis has not previously received significant attention in the literature. This is not to say that ethnogenetic processes have been ignored entirely, but that previous researchers have focused on the Euro-American experience or the syncretic social processes involving both European and indigenous actors in the creation of Hispanic or *mestizo* identities (Deagan, 1973, 1985, 1990a, b, 1996, 1998, 2003; Hoffman, 1994—see also Scarry, 1999; Scarry and Maxham, 2002, for discussion of elite identities in Apalachee).

To the contrary, I know of no study that explicitly considers processes of identity transformation *solely* among indigenous populations in Florida, that is, in the absence of direct Spanish/African cultural input. Although the importance of the Spanish presence cannot be denied, the emphasis in this book is not on these burgeoning *mestizo* communities. Rather, the focus here is on those populations at the edge of the Spanish social sphere, perhaps those less acculturated, willing to be acculturated, or of lower social status and standing. These people were the collective "other."

Perhaps I am alone in proposing that such changes in group identity were occurring (but see Worth, 2006, for discussion of the acculturation concept as applied in the colonial Southeast). After all, ethnonyms did not change during the 17th century and there is no mention of any term that defines a pan-Florida ethnic consciousness, nor is there any direct mention in any sources that such processes were ongoing. Apalachee, Guale, and Timucua ethnonyms continued to be used by the colonial government to organize the provincial structure of the colony and served to differentiate the ethnic composition of villages well after the Franciscan missions had been destroyed and remnant populations had congregated near St. Augustine where they remained for over six decades (Hann, 2003: 101). Of course, even the novice historian would concede that such absence of evidence is not to be overinterpreted. Just because it was not written does not mean it never occurred, and the field of historical archaeology, in part, developed to help fill these historical lacunae. As McGuire (1982: 161) noted, "documentary sources often stress unusual major events, seldom discussing the mundane, everyday processes of social relationships that maintain ethnic boundaries." Indeed, colonial ethnonomy is a complex and problematic area and it certainly cannot be seen as definitive:

> simply because a people use a specific label(s) to distinguish themselves from others does not imply that their spheres of interaction or meaning are restricted to such a designation. Nor does it presume that the ascription, itself, generates a high degree of discrimination or saliency. The relative influence of ethnic ascriptions and their correspondence with other sociocultural phenomena must be established in concrete and particular historical contexts. (Albers and James, 1986: 1)

Importantly, Albers and James (1986) are referring to emic ascriptions and not the etic categorizations characteristic of colonial administrators, or even early ethnographers (Cohen, 1978; Emberling, 1997; Jones, 2002; Levine and Campbell, 1972; Moore, 2001).

However, ethnogenetic theory does not require, or even imply, the complete loss of existing ethnic identities and certainly makes no predictions about how past and present are reified in ascribed nomenclatures, particularly if these labels are objectifying communities in the context of power differentials. As stressed in chapter 3, individuals have identities, not *an* identity, and these identities are segmentary and situational (Cohen, 1978). They are "overlapping sets, groupings which encompass other groupings" (Banks, 1996: 44). That is, an individual's identities are organized into an inclusive hierarchy and they are minimized or stressed as different situations prevail (Barth, 1969; Emberling, 1997; Nagata, 1974). This is the essence of instrumentalist theories of ethnic mobilization in which "individuals are constantly shifting from one to another identity according to context" (Lockwood, 1984: 4). Apalachees traveling to St. Augustine might invoke a completely different discourse set from the one they would use if they were at home, depending on the specific situation they were encountering while away.

Of course, in La Florida, we have scant evidence of any of this. Rather, the record simply reaffirms linguistic differences that have come to represent distinct cultural traditions because each had material analogues visible in the archaeological record. However, to deny that indigenous ethnogenesis was occurring in La Florida is to remove the 17th-century southeastern United States from the hemispheric zeitgeist of colonial transformational processes. Myriad case studies have been developed throughout the Americas that document the process of ethnic tribal emergence in the aftermath of demographic collapse and the resulting novel social and political institutions that represent adaptations to European economies (Albers, 1993, 1996; Albers and James, 1986; Ferguson and Whitehead, 1992; Gonzalez, 1988; Hickerson, 1996; Hill, 1996a). If examples of colonial ethnogenesis are so common and the process so ubiquitous, then why should we assume it was not occurring in La Florida as well, particularly when the same mechanisms of change (demographic collapse, population aggregation, chiefdom decentralization, long-distance migration, a shift to a capitalist market economy) have been so well documented there?

In this chapter, inferences from ethnohistorical and archaeological sources are combined with social theory to bolster the conclusion that this book has sought to establish—that ethnogenesis was occurring among Catholic communities in Spain's La Florida colony. This process was largely invisible to contemporary chroniclers, and yet it is of ultimate importance for understanding modern tribal identities in the southeastern United States. I specifically outline why the period 1600–1650 represented the separation phase of ethnogenesis and reserve commentary about the transition to the liminal phase circa 1650 for chapters 5 and 6. The increase in genetic diversity documented in chapter 1 is interpreted as part of an incipient process of colonial identity transformation that resulted from political decentralization in the aftermath of demographic restructuring and the establishment of a new sociopolitical system with distinct forms of discourse. To understand this argument requires a historical discussion of the nature of ethnic identities.

Tribes and Ethnic Groups

Common to most secondary treatments of the ethnicity literature is a discussion of the historical development of the shift from a concern with "tribe" to ethnic group (Banks, 1996; Emberling, 1997; Jenkins, 1997; Jones, 2002). Although in part reflecting sociological stimuli, including the ubiquity of transnational migration, recent postcolonial histories such as in Africa, and the emergence of global communications and economic systems, this terminological shift was also conceptual, in recognition of the growing divide between theory and empirical ethnographic observation. Tribes were seen as units of investigation, a bounded, monolithic "culture" to be described and categorized as though their development was primarily the result of isolation. The problem for early ethnographers was in defining the boundaries of the tribal group. Objective cultural features such as language distributions or material culture inventories often produced discordant distributional patterns, and economic and political organizations did not covary. This led to the realization that tribal ethnonymy from the ethnographer's perspective was illusory. Reification abounded.

"Tribe" yielded to "ethnic group" in social science research as perceptions of diversity resulting from social isolation yielded to the recognition that interaction was actually partially responsible for the diversity of social forms. This conceptual shift is most often attributed to the work of Barth (1969) and various members of the Manchester School working under Max Gluckman on ethnic diversity in the Copperbelt of Africa (Cohen, 1969; see also Banks, 1996; Eriksen, 2002; Jones, 2002). Although definitions of ethnic groups often focus on fictive kinship and common but nongenetic descent (in opposition to other forms of identity such as class, status, or nationality), the most critical conceptual change effected by Barth and the Manchester School was the focus on interaction as part of the ethnogenetic—that is, ethnic boundary formation—process. Ethnic groups are ephemeral and develop because of interaction. Tribes are static, well-defined, and develop in isolation.

In recognition of these divergent perspectives there are two diametric ways to interpret the initial decline in regional genetic homogeneity among populations in La Florida. The first explicitly invokes the concept of tribe in the postcolonial sense of Morton Fried as "socially bounded cultural groups existing within states or on their peripheries" (Emberling, 1997: 306). Increasing isolation resulted in less permeable population boundaries, and gene flow among communities declined. This approach reifies social and biological populations. The second interpretive framework is more nuanced and is based on Barth's (1969) view of ethnic groups as "categories of ascription and identification . . . organizing interaction between people"; Barth further noted that "stable, persisting, and often vitally important social relations are maintained across such [ethnic] boundaries, and are frequently based precisely on the dichotomized ethnic statuses" (Barth, 1969: 10). Ethnicity, the property ethnic groups share, is "an aspect of a relationship, not a property of a group" (Eriksen, 2002: 12), and as such, the decline in genetic integration among mission communities reflects a loss of identity saliency. Each of these perspectives will be discussed in turn.

The Early Seventeenth Century as a Tribal Zone

In colonial contexts it is difficult to consider the notion of tribe or tribalization without reference to warfare. Both Hill (1996b) and Ferguson and Whitehead (1992) develop ethnogenesis theoretically as a process embroiled with conflict, "a creative adaptation to a general history of violent changes—including demographic collapse, forced relocations, enslavement, ethnic soldiering, ethnocide, and genocide" (Hill, 1996b: 1). Traditional focus on Euro-American violence directed toward indigenous populations, the "conquest" approach to history, has been supplemented (or supplanted) in recent years by a similar concern with the conflict that occurred along multiple dimensions within and among indigenous communities, eschewing the "Indian/European" dichotomy as well as the notion of passive/active responses to colonization and demographic collapse. Rather, attention is focused on the internal social mechanisms of ethnogenesis and the tendency to view Europeans as catalysts and agents of change but not necessarily in a top-down, unilineal direction (see chapters 9 and 10 in Pluckhahn and Ethridge, 2006). The old paradigm of European military supremacy combined with deadly infectious agents leading to widespread passive extirpations is no longer tenable. Instead, comparative ethnographic and historiographic research indicates that the colonial period witnessed myriad indigenous responses to the European presence. The resulting convoluted surface of the adaptive landscape resulted in intense factionalism as cultural systems become internally divided, "struggling to control access to the dominant society's wealth and power...[or] around the issue of how to cope with the dominant society" (Hill, 1996b: 2). Colonial ethnogenesis was therefore an active process of adaptation and negotiation among indigenous communities, and the different strategies adopted were of ultimate importance in determining long-term survivability (Ferguson and Whitehead, 1992; Sider, 1994; Whitehead, 1992; Worth, 2002). Importantly, ethnic factionalism is what leads to tribalization, reducing ethnic-based regional community organizations to postcolonial tribes as part of the "emergence of increasingly separate and distinct native societies, each with increasingly firm and fixed boundaries" (Sider, 1994: 111). These are the tribes created in the sense of Fried (1975), the designs of colonial administrators seeking to "dramatically increase the rigidity of cultural differences . . . as a strategy of control" (Emberling, 1997: 308).

From this theoretical perspective, the Florida frontier can be seen as a tribal zone in which increasingly well-defined ethnic groups (Apalachee, Guale, Timucua) developed distinct strategies for dealing with the European presence and competed among themselves for access to resources and power. Ethnic consciousness intensified as people retreated to their sense of identity in times of significant turmoil. As factionalism ensued and boundary definitions solidified, we might expect a similar decline in between-population biological integration such as that documented during the period 1600–1650. The changes in biological integration documented among communities living in La Florida are a passive signal of the emergence of tribes, with increasingly impermeable and fixed boundaries, in the wake of demographic collapse and the attempted imposition of a Spanish Catholic quotidian existence.

Although this approach satisfactorily explains the biological data in an extremely straightforward manner and positions identity changes as part of a distinctly active indigenous response, historical sources contradict the notion of a La Florida "tribal zone" with Apalachee, Guale, and Timucua components. Foremost, warfare and conflict *among* indigenous groups is an integral component of the tribalization process (Ferguson and Whitehead, 1992). There is absolutely no evidence for increased conflict among Apalachee, Guale, and Timucua communities during the early 17th century, nor is there any evidence that ethnic-linguistic groups were necessarily competitive and adopted independent and different strategies for dealing with the European presence. Although competition among elites for access to Spanish goods was certainly an impetus for conversion to Christianity, there is no reason to believe that different chiefs approached the Spanish in significantly different ways. This does not deny the existence of some competition, the personalization of Spanish-chief interactions, the importance of charisma, or simple self-preservation and promotion. The historical sources, as one-sided as they are, simply lack clear articulation of these points.

In addition, it is known that the Spanish required cooperation from the chiefs and fostered peace among Christian communities as a condition of their receiving friars (Bolton, 1917; Gannon, 1965; Geiger, 1937; Hann, 1988). This point is critical and nearly undeniable. The Catholic communities of Florida which had previously existed in a state of constant, internecine warfare ceased all such activities once the friars arrived. However, violence did not completely disappear from the lives of the converted. It simply shifted in structure and assumed the form of uprisings and rebellions that were local in origin, never involving the entire province or even incorporating elements from widespread ethnic-linguistic groups, a point I return to below.

In addition, the long-term trajectory of ethnic factionalism is ethnogenesis through community *fissioning*, which predicts the development of increasingly rigid Apalachee, Guale, and Timucua identities and tribal boundaries during the 17th century. The exact opposite is documented by the biological data; these political boundaries become more diffuse and permeable throughout the course of the 17th century as long-range migration and resulting gene flow increase in intensity. Because there is no historical or archaeological evidence that precontact boundaries between chiefdoms became less permeable after contact, as expected of the tribal zone model, a different explanatory framework is needed, which I propose is more in line with Hickerson's intended meaning of the separation phase.

The Early Seventeenth-Century Separation Phase

The tribal zone perspective and the notion of the colonial "tribe" are clearly at odds with the nuances of Hickerson's (1996: 70) separation phase of ethnogenesis, which "amounts to the negation or severing of . . . existing group loyalties." The differences between this perspective and that of the violent, factionalized tribal zone are subtle yet important. On one hand, the dissolution of community bonds can be seen as an initial process of tribalization, with intercommunity mechanisms of integration breaking down as competitive and violent ethnic rivalries foment. In this case, the severing of interethnic loyalties precedes an increase in the saliency of ethnic consciousness and a strengthening of boundaries between ethnic groups. However, the separation phase is more passive in form, involving not the increased saliency of ethnic identity but the exact opposite. Ethnic consciousness loses meaning as the factors responsible for initial ethnic boundary formation disappear. Although it is tempting to infer that mate exchange across ethnic boundaries would increase with the decline in ethnic consciousness, this is contrary to the essence of what ethnicity really is. To explain this interpretation, I once again return to the distinction between tribe and ethnic group.

Many scholars of ethnicity have noted the importance of interethnic contact in boundary formation and maintenance, beginning, of course, with Barth (1969), thereby defining ethnic groups in terms of systems of integration rather than culture-bearing entities with historical transcendence. For example, Hudson (1999: 13) defines ethnicity as "cumulative definition through interaction"; for Jones, ethnicity represents a "subjective construction . . . in the process of social interaction" (Jones, 2002: 54) and "a consciousness of identity vis-à-vis other groups" (Jones, 2002: 64). Lockwood (1984: 4) commented, "Ethnicity is peoples in contact . . . [n]o ethnic group can exist in total isolation," while for Eriksen (2002: 1), "Ethnicity emerges and is made relevant through social situations and encounters, and through people's ways of coping with the demands and challenges of life." These writers all follow Barth (1969) in emphasizing the boundaries of ethnic groups rather than their contents and in explaining the formation or maintenance of a boundary through interaction rather than isolation. These conceptualizations of interethnic relations define ethnic groups as components within larger social systems in which information is constantly exchanged across ethnic boundaries despite the maintenance of some social distance. Another important nuance of this perspective is the emphasis on continued contact and interaction for the persistence of social ascriptions of an ethnic nature. As Banks (1996: 32) noted, "ethnic identities do not 'naturally' persist, but need to be maintained." Therefore, when the stimuli that fostered ethnic consciousness and initial mobilization disappear, so too do group boundaries and concurrent practices of integration.

Given this, when Hickerson (1996: 70) speaks of the "severing of . . . existing group loyalties," I interpret this as a decline in ethnic consciousness and ethnic sentiment, not an increase in the saliency of ethnic identity as predicted by the tribal zone model. Biological integration among communities in La Florida does not decline because of increasingly violent and antagonistic interaction between formerly peaceful and integrated communities (the tribalization perspective). Rather, biological integration decreases because systems of intercommunity integration—defined primarily in economic, political, and social terms, but with concomitant biological analogues—lost meaning as European trade items and prestige systems redefined local and regional economies and as the entire fabric of indigenous society was transformed by demographic collapse. The result of this collapse was more sedentary settlement patterns based increasingly on maize agriculture, implementation of a market economy with redistribution channels now funneled regionally through St. Augustine, and ideological conversion to Catholicism. Biological divergence was a passive and unintended result of the decline in integration among groups. Therefore, the initial phase of ethnogenesis through *fusion* involves the removal of those social mechanisms that defined the regional interaction patterns that had resulted in the initial formation of distinct ethnic categories in the past.

Such a theoretical perspective fits well the notions of Anderson (1994a, b), Hally (2006), and Kowalewski (1995) on precontact period chiefdom organization in the southeastern United States. Although I do not presuppose that Mississippian chiefdoms were "ethnic" in character, from a functional perspective an analogy is obvious. Reconstruction of the baseline condition of chiefdom organization is, therefore, a critical component for understanding the separation phase of ethnogenesis in Spanish La Florida.

Establishing a Baseline: Chiefdom Structure in the Southeastern United States

Archaeologists working in the southeastern United States have recognized the complex relationships between chiefdoms and the importance of understanding political organization within a regional framework. The Mississippian period witnessed the rise and fall of chiefdoms over relatively short periods of time, political aggregation during the course of the formation of paramount chiefdoms, the expansion and contraction of chiefdoms as the political standing of specific lineages changed, and the importance of long-distance trade and exchange in the redistribution system central to chieftain authority (Anderson, 1994a, b; Brown et al., 1990; Hally, 2006; Widmer, 1994). In other words, focus on the southeastern "chiefdom" has changed to a focus on Mississippian political economy conceptualized in a broad, regional perspective (Kowalewski, 1995). Whether through chiefdom cycling which entailed the movement of personnel continuously during phases of aggregation (ascendancy) and redistribution (decline) across a broad regional political system (Hally, 2006) or a process of fission and fusion (Blitz, 1999; Hally, 1996), the biological effects in terms of patterns of genetic integration would be evident.

An important aspect of interaction between chiefdoms was incessant, small-scale warfare which fueled the competition for prestige among neighboring polities leading to cycles of aggregation and decentralization among communities (Anderson, 1994a, b; Blitz, 1999; Hally, 2006; Knight, 1986; Steinen, 1992). Therefore, chiefdoms were not isolated, monolithic institutions ("tribes") but existed, in part, because of the very interaction that divided communities into polities (Anderson, 1994a, b; Hally, 2006; Kowalewski, 1995), in parallel with modern perspectives on ethnic groups (see above). Even more crucial for my perspective is the process of cycling or fission and fusion that redistributed the population across the landscape. As Hally (2006: 31) noted, "individual chiefdoms cycled in and out of existence at regular intervals, but the fundamental structural characteristics of Mississippian society remained unaltered. Neither the geographical size, spacing, duration, or number of polities changed appreciably through time." The frequent collapse of paramount chiefdoms resulted in the "periodic movement and re-sorting of large numbers of people among a limited number of polities" (Hally, 2006: 40.) It is important to stress that this argument does not equate individual chiefdoms with different ethnic groups; rather, like ethnic groups, chiefdoms were defined through interaction and opposition, with an exchange of personnel across spatial borders. These archaeological perspectives are remarkably consonant with the level of regional genetic variability presented in chapter 1 and the inference that populations throughout La Florida were less genetically divergent than expected based on recorded levels of sociolinguistic diversity.

Evidence of these systems of exchange has been identified both archaeologically and historically. For example, the Calusa built and maintained a system of canals that implemented trade and redistribution throughout the southern half of the Florida peninsula and directly contributed to Calusa political hegemony throughout much of south Florida (Luer, 1993; Hann, 2003).[1] Another example of the complexity of the transportation network is inferred from the de Soto narratives. According to Hudson et al. (1984), de

Soto's occasional use of native guides suggests that he relied heavily upon existing transportation networks (trails) in his quest for food and fortune (Smith and Hally, 1992), notwithstanding the sometimes arduous nature of the journey (mired in swamps), the extensive buffer zones that separated chiefdoms (DePratter, 1991: 30), and the distinct impression that their guides (many of whom were kidnapped) were less than forthright about village locations and the best routes to take.[2] Similar inferences about indigenous communication networks are suggested by the frequency with which de Soto came upon abandoned villages or well-orchestrated ambushes (see Biedma in Bourne, 1922: 4, 6, 24, 26, 34, 39; Nuñez Cabeza de Vaca in Bandelier, 1905: 127; Elvas in Bourne, 1922: 22, 25, 29, 37, 41, 45, 46; Priestley, 1928: 291; Ranjel in Bourne, 1922: 79, 89, 98), both indicating prescience among the inhabitants (discussed by Steinen and Ritson, 1996). Knowledge of distant political rivals is also suggested by the chroniclers. For example, the Apalachee were well known and feared throughout the Southeast, from Cofitachequi in South Carolina to the western Timucua provinces of central Florida, the Tampa Bay region (see Biedma in Bourne, 1922: 5; Cabeza de Vaca in Bandelier, 1905: 12–13; Elvas in Bourne, 1922: 38; Ranjel in Bourne, 1922: 73), and as far south as the Calusa (Hann, 2003: 24). If Biedma (in Bourne 1922: 11) is to be believed, these interaction spheres covered hundreds of kilometers.

In this sense, the degree of cultural diversity in La Florida does not represent the result of historical isolation and inaction but the exact opposite. Identity differences developed because of the interaction that occurred among different communities that catalyzed social differences, and part of this structure of relationships defined broad regional patterns of mate exchange. There is strong historical and archaeological support for this empirical observation: mechanisms existed that ensured exogamy as a rule of practice.

Protohistoric Period Biological Integration

It is known from ethnographic data on tribal societies in the postcontact period that sociopolitical or ethnolinguistic affiliations have no a priori relationship to patterns of mate exchange, which is often poorly defined and unconstrained by language, polity, or social identity. The phenotypic data presented in chapter 1 indicate that isolation-by-distance was an ever-present feature of southeastern tribal organization and may be reflective of the movement of people within and among polities. For example, Elvas records the presence in Apalachee of a boy who "did not belong to that country, but to one afar in the direction of the sun's rising, from which he had been a long time absent visiting other lands" (Elvas in Bourne, 1922: 50). This is a prime example of the type of intertribal integrative exchange alluded to by Quinn (1993) and Moore (1994a, b, 2001) that are so crucial to the ethnogenetic perspective. Fontaneda's memoir also speaks of his ability to move relatively freely among chiefdoms (learning several languages) with little impunity (True, 1945),[3] as does, apparently, the narrative of Cabeza de Vaca who mentions his capacity as a trader of imported goods (Cabeza de Vaca in Bandelier, 1905: 74–75).[4]

Historical records provide limited but compelling evidence of the degree of contact and integration uniting disparate social communities throughout Florida and Georgia during the protohistoric period. These inferences are based on the few period sources that predate the mid-16th century, and principally derive from the narratives of Cabeza de Vaca, the de Soto entrada, reports from the French occupation of the Florida Atlantic coast, and documents surrounding the Menéndez colonization effort. Focus on these pre-mission period historical sources attempts to produce a more unadulterated "ethnographic present" with full recognition of the limitations of such an approach (Dobyns, 1983; Ramenofsky, 1987). Nonetheless, details gleaned from these sources belie the interactions among distinct political entities and are illustrative of the kinds of exchange networks and power structures within which social identities would manifest. In addition to the biologically homogenizing effects of political cycling and social integrative mechanisms, two other practices that were common during the protohistoric period promoted long-distance mate exchange. Both, we can say with some certainty, stopped very soon after the arrival of the Spanish and both reflect a change in the structure of interactions and therefore contribute to the separation phase of ethnogenesis.

The first practice was critical for political integration and the establishment of paramount chiefdoms but was likely of more minor importance from an evolutionary perspective. Chiefs solidified their status and maintained the sovereignty of their chiefdoms through politically motivated marriages. A number of recorded cases of political marriages have been documented historically. For example, the chief of the Mayaca was the father-in-law of the cacique of the Ais (Hann, 2003: 118), and the cacique of the Ais was the father-in-law of the cacique of the Jeaga (Hann, 2003: 167). The Calusa were particularly effective at using intermarriage as a symbol of political compliance; and when Carlos (the Calusa chief during the mid-16th century) conquered a new chiefdom, it was expected that that chief would send a relative to marry Carlos as a symbol of their tributary status (see Le Moyne in Bennett, 1968: 105; Hann, 2003: 165, 170; Goggin and Sturtevant, 1964; Laudonnière in Bennett, 1975: 111; Reilly, 1981). This practice explains Calusa resistance to Christian monogamy (Goggin and Sturtevant, 1964: 189; Hann, 2003: 176), supports the purported relationship between Carlos and the cacique of Tequesta (Barcia, 1951: 119), and explains the complicated exchanges between Pedro Menéndez and Carlos regarding the marriage of the latter's sister (Doña Antonia) to the *adelantado* (Reilly, 1981). Carlos believed the ties of kinship that this marriage brought about would shift the balance of power in his direction.[5] Le Moyne and Gourgues recorded similar symbolic marriages during the French occupation of the Timucua region (Le Moyne in Bennett, 1968: 102; Gourgues in Bennett, 1968: 213). DePratter (1991: 23–25) notes a number of similar instances among the Apalachee, Timucua, and Guale in which brothers shared authority among neighboring chiefdoms. Of course, it is impossible to know if these relationships were fictive or consanguineal.

The use of women as mechanisms to build political alliances is also suggested in the de Soto chronicles. Although in most cases it is clear that de Soto and his men kidnapped retainers and females (Biedma in Bourne, 1922: 21, 25; Elvas in Bourne, 1922: 25, 39, 45, 48, 62, 81, 83, 85, 95, 117, 138, 143, 144, 158, 178; Ranjel in Bourne, 1922: 58, 61, 71, 78), the texts also suggest that women were sometimes offered by the chiefs of the various villages the Spaniards encountered (Biedma in Bourne, 1922: 16; Ranjel in Bourne, 1922: 113, 115, 116). Ranjel (in Bourne, 1922: 117) indicates that they used these women as concubines "for their foul uses and lewdness," although their primary duties were as servants and load bearers (Elvas in Bourne, 1922: 45, 81). However, on at least one occasion, de Soto was embroiled in a competition between two rival chiefs, both of whom sought to court his favor and tip the balance of power in their direction. One chief, Casqui, offered de Soto a daughter while the other, Pacaha, offered him one of his wives, a sister, and a third woman of high rank (Elvas in Bourne, 1922: 127, 129; Ranjel in Bourne, 1922: 144).[6] This is parallel to the situation in which Pedro Menéndez was involved, as discussed above, and was commonplace throughout the lower southeastern United States, continuing right through to the Creek and Seminole periods (see Boyd, 1952: 118; Hahn, 2004: 43, 95, 118; Oatis, 2008: 211).

Perhaps most interesting is what this indicates about how the invading

Spaniards were viewed by the indigenous populations they encountered. While much resistance was offered in some provinces (Apalachee in particular), other chiefs actively sought the aid of de Soto against their existing political rivals (e.g., Ranjel in Bourne, 1922: 116–122; see also Dye, 1990). Similar behavior was documented during the de Luna entrada in 1559–1561 (Priestley, 1928) in which the Spaniards conspired with the Coosa against their enemies the Napochie (Hudson, 1988; Steinen, 1992). The French also participated in the conflict between the Timucua chiefs in the early 1560s (see Laudonnière in Bennett, 1975: 91, 116, 120; Ribaut in Connor, 1927); and when Menéndez displaced the French he found himself in a similar position with chief Carlos, his internal rival Don Felipe, the Tequesta who were subjects of Carlos, and their regional rivals the Tocobaga (Reilly, 1981: 421). These records indicate that the Spanish challenged the existing power structure and status quo among chiefdoms throughout the Southeast, from northern South Carolina (Cofitachequi) to south Florida (the Calusa) and all parts in between.

Warfare and Biological Integration

The second practice that affected patterns of biological integration was warfare, which was ubiquitous to the extent that a cacique would "carry on warfare . . . with all who share his boundaries" (Garcilaso de la Vega in Varner and Varner, 1951: 488). The causes and consequences of warfare among southeastern chiefdoms have been discussed at length in other treatises on the subject (DePratter, 1991; Dye, 1990, 2002; Gibson, 1974; Hahn, 1995; Larson, 1972; Steinen and Ritson, 1996), and descriptions of how warfare was conducted have been provided by Garcilaso de la Vega (Varner and Varner, 1951: 487–489) and Laudonnière (Laudonnière in Bennett, 1975:) among other secondary treatments. Oatis (2008: 9) characterized southeastern warfare as "a permanent part of a dualistic worldview . . . [in which] [w]arfare was a cyclical, ever-present part of Indian life, as essential to a community's existence as an annual harvest." Because of this, warfare was endemic. It affected all chiefdoms and was primarily of an opportunistic "skirmish and ambush" form rather than a pitched military-style battle. Similar observations have been made among pene-contemporary chiefdom-level societies, and this form of conflict seems entrenched within the political structure and hierarchy of chiefdoms in general (DePratter, 1991; Gibson, 1974; Hally, 2006; Steinen 1992; Steinen and Ritson, 1996).

What is important is that one consequence of internecine, skirmish and ambush-style warfare was the capture of individuals who were enslaved and held within their captor's communities. Although many reasons have been offered to explain the prevalence of warfare during the protohistoric Southeast, and likely multiple causal factors were operative, some have proposed that the capture of prisoners was one of the primary stimuli. DePratter (1991: 52) has argued that captives increased the agricultural capabilities of the enslaving chiefdom, and Garcilaso's account of the conflict between the chiefdoms of Casqui and Pacaha (Capaha to Garcilaso) supports this. After successfully raiding the chief village of Pacaha, Garcilaso noted, "here they [the Indians of Casqui] discovered many of their own people who had been captured and now served as domestics in the lands and fields of the inhabitants of Capaha" (Varner and Varner, 1951: 439). To prevent their escape, Garcilaso reports, they were intentionally maimed and often ransomed (Varner and Varner, 1951: 488). Gibson (1974: 133) also highlighted the central role of capturing prisoners: they added "a basal stratum to social hierarchies, and were used to ward off possible annihilation when the ability of the chiefdom to resist overthrow began to wane."

Mentions of prisoners in foreign territories are scattered throughout the chronicles. For example, Laudonnière offered the following observation about the Timucua: "The kings make wars among themselves, always by surprise attack. They kill every male enemy they can [but] [t]hey spare the enemy women and children, feed them, and retain them permanently among themselves" (Laudonnière in Bennett, 1975: 11). The evolutionary implications of this statement are obvious; the capture of women and children during a battle is mentioned in several locations throughout Laudonnière's narrative (Laudonnière in Bennett, 1975: 85, 91) and continued as a practice in non-Christian communities through the 18th century (Bartram, 1988: 183). Le Moyne's narrative provides an interesting anecdote involving the Calusa and Onatheaqua, a Timucuan-speaking group:

> Oathkaqua, accompanied by a great number of his people, had brought King Calos one of his daughters . . . to give her to him in marriage. When the people of this island found this out, they laid an ambush for Oathkaqua; . . . they captured the bride and all her women, and carried them off to their island. [T]hey are accustomed to marry virgins whom they take in this manner, and to be excessively in love with them. (Le Moyne in Bennett, 1968: 105–106)

Cabeza de Vaca's narrative also provides an alarming and fanciful account, although not specifically from Florida:

> It is a custom of theirs to kill even their own children for the sake of dreams, and the girls when newly born they throw away to be eaten by dogs. The reason why they do it is (as they say) that all the others of that country are their enemies with whom they are always at war, and should they marry their daughters they might multiply so much as to be able to overcome them and reduce them to slavery. Hence they prefer to kill the girls rather than see them give birth to children who would become their foes.
>
> We asked them why they did not wed the girls among themselves. They replied it was bad to marry them to their own kin, and much better to do away with their daughters than to leave them to relatives or to enemies. . . . When they want to get married they buy their wives from enemies. . . . They kill their children and buy those of strangers. (Cabeza de Vaca in Bandelier, 1905: 88–89)

Although we lack detailed data from all chiefdoms included within the Spanish mission system, the historical sources generate a palimpsest that likely applied widely throughout the southeastern United States by virtue of the similar political structures shared throughout the region (and see DePratter, 1991: 26, for a discussion of similar practices among the Powhatans of Virginia). For communities in La Florida, protohistoric sources indicate that the Apalachee were constantly at war with their neighbors on all sides (Hann, 1988: 181–194; Ranjel in Bourne, 1922: 73; Steinen and Ritson, 1996; Sturtevant, 1962: 67), including other future Christian groups such as the western Timucua Yustaga (Barcia, 1951: 77; Cabeza de Vaca in Bandelier, 1905: 22).[7] The Timucua chiefdoms were aligned into confederacies and waged war among themselves as well as with Apalachee (see Bennett, 1968: 103; 1975: 76–77; Milanich, 1999: 48), and the Guale were known to have warred with neighbors to their south (eastern Timucua groups) and to their north (the Orista and Escamaçu) (Geiger, 1937: 67). Therefore, it is likely that the exchange of people was coincident with this style of warfare in the protohistoric Southeast.

Based on these protohistoric accounts we can infer some microevolutionary effect of interchiefdom warfare. However, unlike politically expedient marriages that affected the elite classes only, the capture of personnel through opportunistic ambushes would have targeted the entire community and a larger segment of the population. In addition, while the frequency of conflict between any two neighboring polities may have waxed and waned with political tensions within the upper echelon of society, in general, the opportunities to capture enemies for hostage or integration were more episodic and therefore of greater significance from an evolutionary perspective.

The importance of these observations is apparent when we consider what is known from the Spanish mission period proper. Without question we know that the Spanish successfully brokered peace among Christian communities (Geiger, 1937: 67; Oré 1936: 114–117), those same communities that formed the dataset analyzed in chapter 1. And if interchiefdom warfare decreased among Christian communities, then it is likely that the exchange of personnel across political boundaries also declined, thus resulting in the short-term decrease in biological integration across northern Florida and southern Georgia. This last point must be stressed fervently. When evaluating this explanatory framework it is crucial to draw a distinction between temporal patterns of warfare throughout the Southeast in general versus that experienced among communities within the Spanish mission system (see Hahn, 1995: 15–16, for a discussion of the former). The argument I think is compelling but somewhat counterintuitive: the initial stage of ethnogenetic transformation resulted from a decline in conflict among southeastern communities previously embroiled in internecine warfare, resulting in either a more limited, myopic localized interaction pattern or a more amicable one.

During the separation phase of ethnogenesis in Spanish colonial Florida, indigenous communities were living in an increasingly transitional social environment. Individuals and communities were adjusting to the sweeping changes occurring in their lives that obviated existing mechanisms of social integration; however, the more limited effects of demographic collapse during the early 17th century made more drastic social accommodations unnecessary. As a new system of political discourse emerged, and existing loyalties and rivalries lost meaning, the diverse ethnic communities in La Florida retreated initially to the familiar aspects of their lives. The interaction scale may have been localized and more myopic as the complex systems of allegiance that characterized precontact regional political structure ceased to exist. All of La Florida was being welcomed into the fold of the Catholic Church. An unintended consequence of this devolution was a decline in biological integration, which had previously occurred due to long-term processes of political collapse and regeneration as well as due to internecine warfare. The cessation of warfare among Christian communities is perhaps most integral to this process. This perspective eschews a model of ethnic factionalism and intertribal rivalry (the tribal zone perspective) because there is little evidence for competition among Christian communities after conversion to Christianity. Likewise, there is no evidence, historical or archaeological, for increasingly *impermeable* tribal territorial boundaries between these chiefdoms. Changes in the pattern of biological integration observed during the separation phase of ethnogenesis appear to reflect a passive phenomenon, in concert with theoretical predictions from ethnic identity theory.

Although I have defined both active and passive elements of the separation phase of ethnogenesis, a passive approach to the colonial experience by the indigenous populations of Spanish Florida would not last long. Once demographic collapse resulting from epidemics, slave raiding, and fugitivism exacted its heavy toll on community viability, a more active process of biosocial adaptation would emerge. This signaled the initiation of the liminal period of ethnogenesis in which individuals living in postcollapse societies actively engaged changing hemispheric geopolitics as Spain's control of North America was tested. The specter of England, which dominates U.S. cultural historical discourse, would soon loom large across the southern half of the continent while at the same time Spain's own policies toward, and treatment of, their indigenous allies affirmed the sharp racial quality of sociopolitical discourse during the 17th century.

Five *The Liminal Phase of Ethnogenesis*
Objectification within the Eastern Woodlands Tribal Zone

> It has been common for scholars of the mid-Atlantic coast . . . to view the century before 1670 as a time of limited native movements and scant overland exploration by Europeans. We knew that Iroquoian-speakers . . . were coming in from the northwest . . . . [w]e heard . . . that the man Powhatan had single-handedly built up his paramount chiefdom to cover most of eastern Virginia. We had Smith's references to the Jamestown colony sending parties . . . in search of the survivors of the "Lost Colony." We also saw that an English expedition tried unsuccessfully to establish a trading relationship with the Tuscaroras . . . in 1650 . . . . *But that is all*. [However] the curtain began to go up around 1650 not 1670 (emphasis added).
> —Helen C. Rountree, "Trouble Coming Southward" (2002: 65–66)

The dramatic changes in population structure that occurred during the late mission period (post-1650) suggest a significant response to changing sociopolitical circumstances by the native populations of Florida and Georgia. I have interpreted the biological data as reflecting the emergence of a panethnic consciousness among La Florida's indigenous communities. In this chapter I outline the changes that occurred during the late mission period that differentiated the experiences of early and late 17th-century mission populations such that the liminal phase of ethnogenesis was initiated. I focus upon two distinct yet complementary aspects of ethnogenesis, with particular emphasis on ethnic boundary shifts (Barth, 1969) and further consideration of the concept of a tribal zone, tribalization, and tribe-state interactions (sensu Ferguson and Whitehead, 1992; Wolf, 1982). In defining ethnicity as "a boundary from within, maintained by the socialization process, and a boundary from without established by the process of intergroup relations," Isajiw (1974: 122) highlights two critical components of the ethnogenetic process referred to as objective (from without) and subjective (from within) elements. As Eriksen noted, "ethnic identities are neither ascribed nor achieved: they are both" (Eriksen, 2002: 56), a distinction reflected by Jenkins (1986: 177) by the terms *group identification* versus *categorization*; the former is an internal process, the latter "relates to the ability of one group ... to impose its categories of ascription upon another set of people" (Jenkins, 1986: 177).

Objective elements of ethnogenesis refer to the extraregional stimuli that etically define diverse communities of people as singular, thus ignoring any internal ethnic or social differences that factored little in structuring interaction between objectified and objectifying communities. It is the objectification of communities of people by others that fosters ethnic mobilization because "characterization ... affects in significant ways the social experience(s) of the categorized" (Jenkins, 1997: 53). In this way, social boundaries appear between the objectified and objectifying communities by virtue of this structure of relationships. However, objectification can also cause a loss of ethnic differentiation among those communities subject to the objectifying forces; boundaries between these ethnic communities dissolve due to recognition of a common political basis of mobilization and, by extension, a shared historical fate. That is, when diverse ethnic groups are treated similarly by a dominant political institution, internal ethnic differences may lose saliency, or, alternatively, diverse ethnic communities may construct a higher order identity for the express purpose of interacting with the objectifying political agent. The former represents a typical case of assimilative ethnogenesis; the latter is more developed within the modern identities literature in the form of panethnic, that is, pan-Asian, pan-Latino, pan-Indian, movements (Espiritu, 1992; Tefft, 1999).

The subjective component of ethnogenesis refers to the ways in which people respond to objectification by developing emic definitions of identity which are codified culturally through the adoption of an ideology and mythology of descent, with accompanying symbolic reification. The latter (subjective) generally follows the former (objective) because, as noted by Horowitz, "self-definitions and other-definitions do not necessarily adjust at the same rate, simultaneously producing a new identity. Especially if the new identity is wider than the old, other-definitions are likely to be 'ahead of' self-definitions for some time" (Horowitz, 1975: 131). Therefore, when faced with intense objectification "from above or beyond," individuals minimize their internal differences and highlight their commonalities. They broaden the basis of affiliation which leads to the emic mobilization of ethnicity through active manipulation of material symbols of identity to meet some political or economic end (situational or circumstantial aspects of ethnicity); to satisfy a need for rootedness, a cultural, psychological, or ideological end (primordialism); or by default, through development of a new habitus (practice theory) in which "sensations of ethnic affinity are founded on common life experiences that generate similar habitual dispositions" that give "members of an ethnic cohort their sense of being both familiar and familial" (Bentley, 1987: 32, 33).

Because both objective and subjective mechanisms are critical to the ethnogenetic process, I expect to find elements of both in the historical record that supplement the biological data presented in chapter 1. That is, if the hypothesized linkage between identity, biological patterns, and material culture developed in chapters 2 and 3 is valid, then subsidiary data should not only reflect the myriad stimuli that contribute to the ethnogenetic process but also affirm the specific timing at which these transitions occurred. Once again, I reiterate that no direct evidence exists in the historical record that ethnogenesis was occurring among La Florida's indigenous populations. Having said this, for communities in La Florida it is surprisingly easy to identify objectifying processes. These are events, policies, or practices that treated the residents of the Spanish Catholic missions as monolithic, that failed to appreciate or recognize internal ethnic differences, and that fostered a feeling of shared persecution regardless of tribal, linguistic, or ethnic status. Such interactions are easy to envision in La Florida, with both local and regional components. The former represent Spanish policies of population reorganization and aggregation, while the latter refers to the unique English New World economic strategy and its dramatic repercussions for eastern North America. These objective elements of ethnogenesis are largely responsible for the *timing* of identity transformation—the transition from the separation to the liminal phase of the ethnogenetic process.

On the other hand, identifying subjective elements of ethnogenesis is much more difficult because these are the types of nuanced behaviors, feelings, and dispositions that are unlikely to be recorded in the coarse pages of history. However, residues of these emic sentiments are evident both archaeologically, bioarchaeologically, and historically. They elicit an image of an increasingly homogenous life experience that differed significantly for each succeeding generation born into the mission communities. These shared life experiences engender those feelings of belonging that are so crucial for ethnic mobilization to persist, regardless of the external mechanisms that may have fostered initial ethnic mobilization. I turn to these historical data next, beginning with discussion of the objective components of ethnic identity formation during the latter half of the 17th century and reserving inferences about emic processes for chapter 6.

Objective Aspects of Ethnogenesis and the Eastern Woodlands Tribal Zone

In the New World there is no doubt that Europeans were an objectifying element, and the Spanish in Florida are no exception. Despite continued use of specific tribal/linguistic ethnonyms throughout the 16th, 17th, and 18th centuries (Hann, 2003: 100–101), there is historical evidence that the Spanish were quite insensitive to the nuances of ethnic variation within their colony (discussed broadly in Waselkov and Cottier, 1984). In fact, with the exception of the Apalachee, the tribal ethnonyms (Guale, Timucua) used by historians and archaeologists today are Spanish constructions and broadly conceived (see chapter 3, also Hann, 1996a).[1] During the 17th century such labels came to assume a geographic and provincial character, although their use as ethnic labels did continue. Nonetheless, the Spanish conception of their colonial empire was quite distinct from that of the French and English. Rather than viewing the Indians as chattel or as impediments to progress, the Spanish viewed indigenous communities as comprising a distinct position within the social and political sphere that was divided into two components, the *república de españoles* and the *república de indios*. These two separate entities "were to be united in allegiance to the Crown and obedience to the 'law of God;' otherwise they were intended to stay strictly apart" (Bushnell, 2006: 198). This compartmentalization in effect created a superordinate identity that may have had little meaning to the Apalachee, Guale, and Timucua but which flavored their interactions with the Spanish. In other words, the

Republic of Indians assumed a panethnic, superordinate identity category with initial top-down significance. In particular, this structuring classification system had a distinct racial and class basis, making it all too easy to objectify "Indians" of various ethnic groups. As McAlister (1984: 395) noted, speaking more specifically about Andean and Mesoamerican contexts, "they were smaller and less robust than Europeans. Their hair was straight and coarse, and their beards grew sparsely. Their color ran from medium to dark brown, a shade that the Spaniards described as *aindiado*." In McAlister's analysis (1984), the Florida communities would have all been *indios de pueblo*, sedentary populations subject to taxation and labor conscription for whom the patriarchal quality of *miserables* would have been applied.

The homogenous perception of the "Republic of Indians" was most clearly demonstrated through the strategy of population aggregation and reorganization that the Spanish implemented as demographic collapse and fugitivism emptied the missions. Indigenous populations were treated like pawns in a complex game of demographic restructuring, initially in the form of local aggregation and *doctrina* consolidation for the purposes of easing the duties of the friars (figure 5.1)(Gannon, 1965; Geiger, 1937; Worth, 1998b). Eventually, however, once population size declined by over 90 percent in the eastern coastal and St. Johns River provinces, more dramatic restructuring occurred that included the forced resettlement of entire communities, the forced aggregation of disparate communities into a single mission center often at a new location, and the resettlement of non-Christian populations within the midst of the Christian communities, even when these same populations were contributing to internal unrest in other parts of La Florida (see Hann, 1996a; Worth, 1995, 1998b). These practices indicate that ethnic nuance figured little in dealings between Spaniard and Indian. The Spanish needed a populace distributed in certain places to perform specific tasks vital to the functioning of the colony (such as manning a ferry crossing), and they needed unmarried males who could contribute to the *repartimiento* system in St. Augustine.

Consider, for example, the Chisca who were pressed to resettle key Timucua missions in the St. Johns River region during the latter half of the 17th century. This is particularly intriguing because the Chisca were a perennial problem for the Spanish; they had been raiding Spanish missions intermittently since 1618 (Hahn, 2004: 17; Worth, 1995, 1998b) and were directly responsible for inciting the Apalachee revolt of 1647 (Hann, 1988: 17–18) and the 1675 revolts in the Chacato missions of San Nicolás de Tolentino and

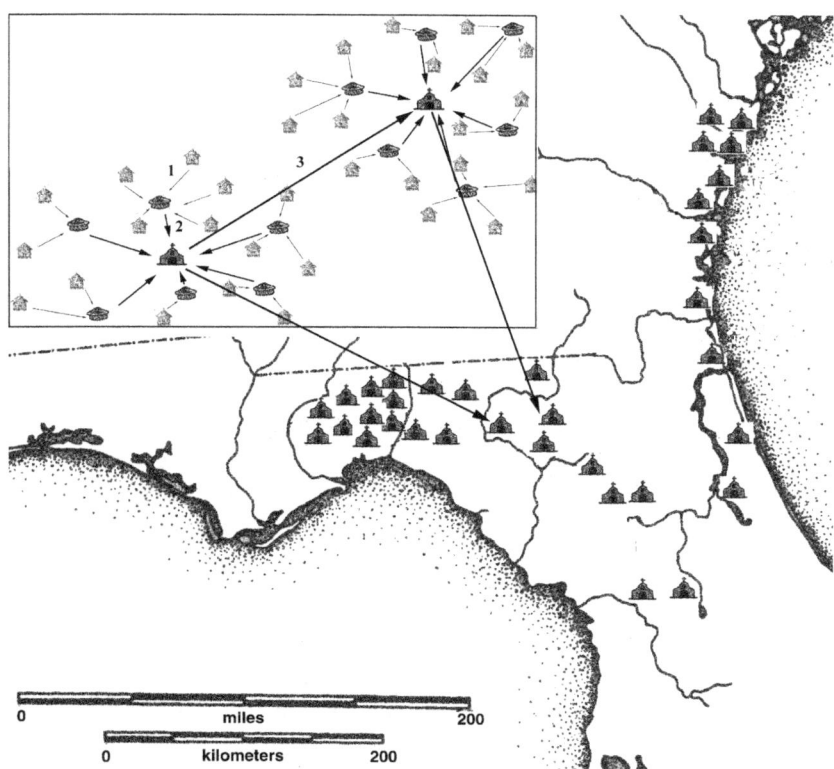

Figure 5.1. Idealized representation of the process of population aggregation during the 17th century. The hierarchical process began with the aggregation of local households to the seat of the primary village (1), the aggregation of villages to the seat of the mission *doctrina* (2), and ultimately resulted in the aggregation of *doctrinas* (3).

San Carlos de Yatcatani (Hann, 1988: 184). That some Chisca were invited to settle within La Florida while at the same time other Chisca were raiding Apalachee (Hann, 1988: 182–183), Timucua (Worth, 1992: 154), and Guale missions (Worth, 1995: 45, 52) belies two important details about Spanish-Indian interactions. First, my primary point, that the Spanish reified the Republic of Indians and by doing so created a superordinate identity category—"Indian." Second, ethnonyms in the colonial South have no relational parallel to political units with any degree of homogeneity, thus allowing ethnic Chisca to simultaneously raid and settle Spanish missions at the band level. For example, Chisca were used as interpreters to interrogate four Westo prisoners captured after the 1680 raid on Guale (Worth, 1995: 16) and also conspired with the Spanish in an attack on the Chattahoochee River communities in the mid-1680s (Kelton, 2007: 135). The Chisca were enemies of the Westo and Apalachicola, yet they collaborated at times with both groups during raids on Spanish Florida (Hahn, 2004: 46; Worth, 1995: 26, 45, 109).

Similar behavior was also evident among the Yamassee and the Chacato (Worth, 1995: 17). The Yamassee in particular have a complex history of Spanish relations. After 1650 their presence was noted throughout La Florida, particularly in Guale and Apalachee while also residing partly in the Georgia interior. They defected from Florida during the 1680s, established a trading relationship with the Scots at Stuart's Town, and then when this colony was destroyed, with the Carolinians at Charles Town when they then began attacking the Spanish missions (Hahn, 2004; Oatis, 2008). Ultimately, the Yamassee became disenfranchised with the English Carolinian traders and were betrayed by the Cherokee during the Yamassee War of 1715; after this they once again returned to seek the protection of the Spanish (Covington, 1968; Hahn, 2002, 2004; Kelton, 2007; Worth, 1995) until they were basically destroyed by the South Carolinians and their indigenous allies (see Oatis, 2008).

That the Spanish sometimes failed to appreciate cultural or ethnic nuance is also reflected in *repartimiento* policies, which were not only blind to the ethnic diversity within the colony but also ignored the preexisting status structure within indigenous communities. Such oversight was directly implicated as the cause of the Timucua rebellion in 1656 (Milanich, 1999; Worth, 1998b). As the eastern provinces suffered demographic collapse during the earlier decades of the 17th century, the Spanish government in St. Augustine continued to look west toward Yustaga and Apalachee to fulfill labor quotas (Bushnell, 1978: 418). When these sources were not sufficient, pagan populations, such as the Yamassee and Chisca, were allowed to settle within the boundaries of the mission system to replenish the much-needed labor supply (Worth, 1995, 1998b). Faith was clearly secondary to economic motives. These types of decisions on the part of the Spanish reflect a basic dichotomy between "Indian" and "Spaniard" that certainly had an objectifying element to it.

As egregious as some of these examples may be, particularly the Chisca, Spanish objectification of the Republic of Indians does not provide a reasonable explanation for why the liminal stage began sometime around 1650. To fully develop this aspect of the colonial ethnogenetic process, the scope of inquiry must be broadened considerably to include much of eastern North

America, and additional actors, primarily the English but also to some extent the French,[2] must enter the historical stage. The second half of the 17th century witnessed a reemergence of the intense competition between European colonial powers for New World dominance, which further objectified and tribalized indigenous populations in Florida and throughout the Eastern Woodlands. While the expansion of one state can effect significant sociopolitical disruption, competition among multiple expanding states creates even more dynamic colonial contexts (Ferguson and Whitehead, 1992; Whitehead, 1992), and this, I propose, is paramount to understanding the timing of ethnogenetic changes in La Florida.

English Contributions to La Florida Ethnogenesis

The English presence in the southeastern United States begins with Roanoke in 1585 (a political but perhaps not epidemiological nonevent; see Kelton, 2007: 77–78; Kupperman, 2007; Miller, 2000; Stick, 1983) and Jamestown in 1607. However, the early years of the Virginians were marked by hardship (Kupperman, 1979; Stahle et al., 1998) followed by several decades of conflict with the Powhatan confederacy (Gleach, 1997; Rountree, 1993; Rountree and Turner, 1994). Violence escalated in 1622 and again in 1644 but resolved with finality in 1646 with the defeat of the Powhatans, thus paving the way for the English to expand trading routes west and south. It was only after the defeat of the Powhatans and the restoration of the English monarchy in 1660 that their bid for New World supremacy could commence,[3] and it is this mid-century date that helps explain the timing of ethnogenetic change in Florida. As commented upon by Bolton and Ross (and as this chapter's opening epigraph indicates), "the shadow of Jamestown soon projected itself into the Carolina-Georgia back country" (Bolton and Ross, 1925: 23).

This emphasis on English history is critical to understanding the transition to the liminal phase primarily because of the different English and Spanish approaches to the indigenous populations within the colonies which further emphasized ethnic divisions and catalyzed the ethnogenesis process. Unlike the English who saw the Powhatans as competitors with no place in their world (Rountree, 1993), the Spanish saw their vassals as comprising an important, though low and distinct, position in their society (Bushnell, 2006; Perdue and Green, 2001: 58). The greatest disparity was both economic and ideological. The Spanish never implemented a plantation system[4] nor did they transfer to Florida the *encomienda* system which had previously provided the early 16th-century impetus for Spanish slaving along Florida's coasts.[5] Importantly, the skin and fur trade, while present, was not as developed in the 17th century Southeast as it was farther north (see Smith, 1987; Waselkov, 1989). Economic disadvantage abounded and *repartimiento* remained the primary means of indigenous abuse within the broader goal of wealth extraction (Bushnell, 1981, 1994; Lyon, 1990; Matter, 1973). While secular and private enterprises struggled (Bushnell, 1978; Worth, 1998a), the dominance of the Franciscan initiative within the broader structure of the colony made saving souls the overall priority. The conflict between soldier and priest was omnipresent in Florida and had a lasting effect. As noted by Chatelain, "the religious policy was in certain other respects detrimental to the welfare of Spanish Florida, in that it involved the curtailment of farming, fur trading, and other economic activities which might have resulted in bringing a degree of prosperity to the inhabitants" (Chatelain, 1941: 38). The Franciscans also actively sought to keep European weapons out of the hands of their converts (Chatelain, 1941: 38). And, although their power waxed and waned throughout the 17th century, the Crown often acquiesced to the wishes of the friars.

The English had no initial religious imperative and saw a quick profit and land ownership as trumping all else. The decade following the defeat of the Powhatans would be spent reconnoitering Virginia and taking advantage of the quick profits to be earned farming tobacco on the Virginia coastal plain, with continued involvement with the fur trade in the north (Gleach, 1997; Rountree, 1993; Rountree and Turner, 1994). However, a decline in tobacco prices mid-century, combined with a shift away from the northern fur trade (the Maryland colony now interceded),[6] enhanced the Virginians' interest in the slave export business (Covington, 1967; Gallay, 2002; Rountree, 1993), while continued focus on tobacco and rice as agricultural products required ever greater contributions of slave labor. The largest purchasers of Indian slaves[7] were the English colonists in the Caribbean, whose colonies also posed a direct threat to Spanish sovereignty.[8] In 1670, Charles Town was founded by businessmen from Barbados on the coast of South Carolina, effectively challenging Spanish control over the southern half of the continent. The Carolinians would continue the commercial development of the Virginians, inheriting from their Barbadian roots a culture of mercantilism that was "exploitative and materialistic" (Edgar, 1998: 38).

It was this hedonistic, wealth-centered fervor that clearly defined the limits of Spanish and English control (Hahn, 2004). Indigenous populations that were not English allies or serving the English as slave raiders were targeted in raids (Bowne, 2005; Covington, 1967; Ethridge, 1984, 2006; Gallay, 2002; Hahn, 2004), although the opening of the Middle Passage[9] and the poor result from using indigenous labor sources ultimately led to the intensification of African slave importation.[10] At the same time, English traders pressed to open the southern frontier and expand the deerskin trade. All of this occurred after 1650, after the defeat of the Powhatans, and at the same time that Westo slave raiders appeared at the edges of Spanish Florida. The Westo had a chilling psychological effect on the Florida mission populations through their capacity as ethnic soldiers, an economic strategy the Westo themselves unwittingly sowed to their own destruction.

Post-Mid-Century Slave Raids and the Escalation of Violence in La Florida

Of paramount importance during the early years of English economic expansionism was an enigmatic group called the Westo. The historical ascendance of the Westo has been the subject of debate for nearly a century (Crane, 1918, 1919, 1956; Juricek, 1964; Swanton, 1922); however, Bowne's latest treatment of the literature concurs with Crane (1956) and Wright (1981) in linking the Westo to a group of Erie displaced by the Five Nations Iroquois during the Beaver Wars who immigrated to the James River region of Virginia, with firearms in hand, around 1656 (Bowne, 2005, 2006). Westo attacks on the interior populations of Georgia and South Carolina fed the English thirst for slaves and forced indigenous peoples to loosely ally themselves with the English or Spanish (or at times both) for protection (Bowne, 2005; Hahn, 2004; Smith, 1987).

The first mention of a Westo presence in the South was in 1659 (note the date). They were located 80 leagues north of Apalachee (in Tama) where they were conducting raids on these interior populations with great success (Hahn, 2000: 67–68). In 1661, some 500–2,000 Westo descended on the Spanish provinces, attacking Santo Domingo de Talaxe and San Joseph de Sapala, both in Guale province (Worth, 1995: 15–18). In 1662, they attacked the pagan town of Huyache, and the following year they resettled along the Savannah River (Bowne, 2005: 78) and continued to launch attacks along the South Carolina coast (the province of Escamaçu) throughout the remainder of the decade (Worth, 1995: 20–21). Finally, in 1680 a group of Westo, Uchise (Creek or Coweta), and Chiluque returned to the province of Guale and attacked settlements on St. Simons and St. Catherines Islands, causing the further contraction of populations along the Georgia coast (Worth, 1995: 25–26; Swanton, 1922: 91).[11] The armed Guale were able to offer some response; however, the damage had been done and the attackers retreated.

The eventual betrayal of the Westo by the Carolinians did not ameliorate the antagonistic conditions of the southern frontier. In fact, the appearance of the Westo initiated a protracted five-decade escalation of conflict between England and Spain that ended with the destruction of the Spanish missions in 1706. The role of the Westo as ethnic soldiers and slave raiders was easily and quickly filled by other groups, and violence continued to escalate throughout the 17th century and the duration of Queen Anne's War. For example, Yamassee raiders attacked Santa Catalina de Ahoica in 1685 (Covington, 1968: 10; Worth, 1995: 45), destroyed San Juan de Guacara (Timucua province) in 1691 (Hann, 1996a: 265–266), and attacked the Timucua mission of Santa Fé in 1702 (Hann, 1996a: 293). That the Yamassee had recently emigrated from the Florida missions is illustrative of the complex interethnic discourses during the 17th century (see Hahn, 2004).[12] Other groups such as the Chisca also intensified their attacks against the Spanish missions, targeting Guale and Timucua soon after mid-century (Hann, 1996a: 238; Worth, 1992: 153–154, 1998b: 18–21), inciting a 1675 revolt among the newly founded Chacato missions north of Apalachee (Hann, 1988: 184), and launching a direct attack against three villages near Apalachee province in 1676 (Hann, 1988: 185). Kelton (2007: 114–115) noted the presence of ethnic Tuscaroras, Ocaneechies, Shawnees, and Tomahitans, all of whom participated in raids on the Spanish missions during this time period.

During the 1680s, Spanish rapport with the Apalachicola declined and the next two decades would witness an intermittent series of Apalachicola raids against the Catholics, with subsequent reprisals (Hahn, 2004: 34–51; Hann, 1988: 264–283). With the onset of Queen Anne's War, hostilities further intensified between England and Spain resulting in a series of raids directed by Colonel James Moore, one of the Goose Creek men from Carolina, and Apalachicola allies. The assault on mission Santa Fé in 1702 by a force of Yamassee and Apalachicola instigated an Apalachee, Timucua, and Chacato retaliatory expedition that same year. Shortly thereafter, James Moore and allies assaulted the Georgia coast and St. Augustine. He returned in 1704 and 1705 and destroyed Apalachee and much of western Timucua, delivering the final blow to the remnant mission populations in 1706 (Boyd et al., 1951; Covington, 1972; Hann, 1988, 1996a); at this point the remaining Spanish loyalists retreated to the safe haven of St. Augustine and the Castillo de San Marcos.

The Effect of Slave Raiding on Colonial Ethnic Alignments

That the Westo assaults on La Florida were of limited success is in some ways irrelevant. Their effects on the climate of the southeastern United States were palpable. When the Westo arrived in Virginia in 1656, very few Indians had firearms. By 1680, as a direct result of Westo depredations, much of the southern frontier was armed, including some of the Spanish mission Indians. Competitor groups, such as the Savannah and Ocaneechee, had armed themselves as well (for protection from the Westo), and these same groups also assumed a predatory existence and replaced the Westo after their defeat (Bowne, 2000, 2005; Crane, 1956). After 1685, Yamassee and Apalachicola increasingly participated in armed aggression against the Spanish missions, with devastating effect (Hahn, 2004; Oatis, 2008). The question is, what emotional or psychological toll was effected by the escalation of violence during the latter half of the 17th century?

It is safe to say that the slave raiders initiated a climate of fear in Spanish Florida as individual communities suffered direct assaults by the raiders or welcomed migrants from central Georgia who were escaping the violence. For example, in 1675 Bishop Gabriel Díaz Vara Calderón described the Westo as "the numerous nation of the Chichimecos, heathen, so savage and cruel that their only concern is to assault villages, Christian and heathen, taking lives and sparing neither age, sex nor estate, roasting and eating the victims" (Calderón in Wenhold, 1936: 11). References to the cannibalistic nature of the Westo are legion (see Cheves, 1897: 166–167, 194, 200–201, 238–239, 334; Crane, 1956: 12; Swanton, 1922: 66, 67, 68; Wenhold, 1936: 11) and as Eric Bowne (2005: 70) noted, the fear was well founded: "They were a foreign group unfamiliar to southern Indians, armed with loud and deadly European weapons that inflicted wounds beyond the ken of native healers. In addition, they had come into the region with the purpose of raiding and slaving for the better part of their livelihood." Importantly, firearms did appear to heighten their perceived advantage as they were known to have "guns and powder and shot" (Cheves, 1897: 334, 194) and were described as "people so addicted to arms . . . [that they] prey upon people, whom they either steal or force away" (John Lederer in Alvord and Bidgood, 1912: 160). As Ethridge (1984: 21) noted, the fear of enslavement was powerful during this time period: "A simple hunting excursion, a walk to visit a nearby relative, or even a step outside the village to collect firewood could, and often did, end in shackles at the public auction in Charles Town. This fear became a morbid preoccupation to the Indians, a terrible fixation on a terrible reality."

The issue of firearms is a difficult one filled with contradictions and misconceptions (Given, 1994; Malone, 1991).[13] Perhaps the most obvious is the legal status of the firearms trade. While it is true that the Spanish prohibited the sale of firearms to Native Americans (Bushnell, 1981: 29), the English were not rampant peddlers of firearms, at least legally. In fact, the sale of firearms was variously prohibited and allowed during the 17th century (Bellesiles, 1998: 578; Hahn, 1995: 66; Kelton, 2007: 110–111). The difference, however, is that English traders circumvented or ignored these rules because of demand and because it increased their profit by improving the hunting and raiding efficiency of the indigenous traders. To the contrary, most groups allied with the Spanish were poorly armed, particularly during the early to mid-17th century, and this put them at a distinct disadvantage. While firearms-related materials were excavated from numerous Spanish mission sites[14] and the historical use of firearms by Spanish mission Indians is also recorded (Bushnell, 1981: 9, 41, 51; Hann, 1988: 186; Worth, 1992: 244–246, 1995: 31), guns were never as numerous or widespread in Florida as they were in the Georgia and Carolina backcountry.[15] The illegal trade of firearms was more limited in La Florida, which may reflect concerted Spanish efforts to keep their distribution contained.[16] In addition, those weapons that were present in Spanish Florida were of an inferior type, as the accounting of arms in St. Augustine attests.[17] In particular, the Spanish relied on matchlocks, which were technologically inferior to the flintlocks that would come to replace them.[18] Therefore, the dynamic between English and Spanish allied tribes was not necessarily simply one of guns versus bows but rather a matter of scale. The non-Christian communities had better firearms and more of them, and they used these to distinct advantage as the English were quick to remind them (e.g., Fitch, 1916: 181).

That the Westo and other predatory slave raiders effected a psychological toll was well represented by the actions of the Guale attacked in 1680. They clearly were eager and anxious to retreat farther south to be closer to the protections of St. Augustine, and they even threatened to commit mass suicide if forced to return to St. Catherines Island (Worth, 1995: 33). This is quite an insight into the culture of fear that pervaded the Spanish colony. Hahn (2000: 69) interprets the sudden change of sentiments (in 1659) among Apalachee and the friars in that province toward the building of a fort at San Luis as a direct result of the rumors and fears of Westo attacks on the missions. And surely, the refugees who began pouring into Guale, Timucua, and Apalachee shortly thereafter[19] told of the horrors experienced at the hands of the Westo (e.g., Alonso Solana Declaration of 1675 in Reding, 1935: 174–175), and the *repartimiento* draft that brought members of many different ethnic groups to St. Augustine would have provided a salient vector of information transfer. All would have known of the changes in the political climate after 1659 and they would have been targets of these English-sponsored raids. All of the Spanish-allied populations were objectified by the Westo and other slave raiders of the colonial Southeast.

However, conflict along the southern frontier should not be conceived as too top down, and as such the use of the term "allies" may be an overstatement. English colonists saw very little loyalty in their trading relationships with the various indigenous communities throughout the South (Oatis, 2008). It would be entirely incorrect to think of these peoples as "English allied." However, ideological differences may have significantly defined the interaction dynamic *among* indigenous communities that roughly tracked along lines of faith. Because the English had little initial interest in conversion to Christianity, the Spanish populations of La Florida may have been viewed by Englishmen and Creeks alike as "Catholic." Christian communities were specifically targeted during late 17th-century slave raids and Christian symbols were intentionally violated during the more serious assaults, suggesting that one goal of the Creek and Yamassee raiders may have been to "cleanse the land of its Catholic scourge" (Hahn, 2004: 55). Indeed, the behavior of the slave raiders during this time period suggests a clear objectification of Spanish Catholic Indians. As noted by Hahn (2004: 60–61), "That the Ocheses [Creeks] chose to kill a Timucuan man as a sign of revenge against the Apalachees and Chacatos was no accident, for evidence suggests that the Ocheses may have considered all Christian Indians as a single, monolithic threat."

Ethnogenesis and the Eastern Woodlands Tribal Zone

The violent encounters detailed above further emphasize the objective elements of ethnogenesis, consistent with Barth's emphasis on ethnic boundaries as the focus of investigation. Barth noted the "tendency towards canalization and standardization of interaction and the emergence of boundaries which maintain and generate ethnic diversity *within larger, encompassing social systems*" (emphasis added) (Barth, 1969: 18). During the latter half of the 17th century the "conditions making up a particular context of interaction" (Jones, 2002: 96) had changed due to the English bid for New World supremacy resulting in the "imposition of new categories of difference" (Jones, 2002: 101). Such changes in the structure of intergroup interaction patterns redefined ethnic boundaries, which I propose assumed a broader, regional structure. That such changes in the scale of interaction can effect ethnogenesis is consistent with Horowitz's view of ethnic boundaries:

> two types of variable seem to be most influential in shaping and altering group boundaries. The first is contact with ethnic strangers perceived as possessing varying degrees of likeness and difference. The second is the size and importance of the political unit within which groups find themselves. These two are, of course, related. Political boundaries *tend to set the dimensions of the field* within which group contact occurs. That contact, in turn, renders it necessary for groups to sort out affinities and disparities" (emphasis added). (Horowitz, 1975: 121)

A more perfect quote could not be found to describe the situation in the late 17th-century southeastern United States. Ethnic diversity among La Florida's indigenous communities was minimized during the process of assuming a panethnic, assimilative form, while an intensification of an ethnic boundary, or boundaries, was occurring as new interaction systems, primarily antagonistic although not completely so, emerged and fostered ethnic realignments with populations farther north.

The La Florida Tribal Zone

This focus on interaction scale is complemented by similar concern with conflict, and it is useful at this point to revisit the concept of a "tribal zone" as previously discussed in chapter 4. Although I ultimately rejected this model as a means of conceptualizing ethnogenetic changes during the early mission period, it provides much greater clarity here. Ferguson and Whitehead (1992: 3) defined a tribal zone as "that area continuously affected by the proximity of a state, *but not under state administration*" (emphasis added), which emphasizes an important distinction related to regional scales of interaction.[20] Hill (1996b: 7–8) succinctly summarized the concept of the tribal zone as "a sphere of interaction in which state-level expansion reduces multilingual, multicultural regional networks to territorially discrete, culturally and linguistically homogeneous 'tribes.'" This is an apt description of the events I propose were ongoing during the second half of the 17th century.

In the previous chapter I rejected the notion of a La Florida "tribal zone" on the grounds of content; there was no evidence for increased conflict among La Florida's converted populations or for increasingly impermeable social boundaries, although one could argue that the record of indigenous revolts do represent "wars of resistance and rebellion" although in a slightly different form than originally intended (see Ferguson and Whitehead, 1992: 18). Rather, the trajectory of biological changes presented in chapter 1 suggests the opposite; the overall trend is for increasingly permeable social boundaries and widespread migration. By expanding the interaction scale during the latter half of the 17th century, however, the limitations defined in terms of relational proximity are no longer evident. And, in fact, the objectification of Spanish-allied populations, both by the Spanish and the English and their indigenous allies, is entirely consistent with a broad corpus of research on tribe-state interactions and ethnic boundary transitions (see Comaroff, 1987; Emberling, 1997) in which "indigenous and Afro-American peoples . . . found themselves internally fragmented and divided against each other in the service of colonial domination" (Hill, 1996b: 5).

In this sense, the fact that the English provisioned with firearms those populations loyal to their interests helped "tribalize" Eastern Woodlands groups and established a dynamic of raiding and conflict distinctly different from that which existed prior to European contact (Bowne, 2005; Hahn, 2004; Sider, 1994; Worth, 2002). By the same token, by enacting a different economic mode, the Spanish tribalized indigenous communities within their province. In the Eastern Woodlands some populations became slave raiders and existed in a predatory form of organization, while others became the targets of those raids. By aligning themselves with specific European economic modes, groups were effectively, although passively, taking sides. From this perspective the tribal zone was not just La Florida but much of the southeastern United States and I propose it had three key components: (1) the Spanish province of La Florida defined by the limits of the mission frontier, (2) the English settlements of the Carolinas and Virginia and their respective "settlement Indians," and (3) the interior hinterland removed from direct contact with any European power. Ethridge (2006) has proposed a similar, broadly defined, eastern U.S. interaction sphere which she called a "shatter zone." Populations living in the interior enjoyed the greatest flexibility of political organization and economic strategy, which resulted in a considerable power imbalance (see Hahn, 2004). And it was here that many of the postcolonial tribes with any degree of temporal endurance emerged. Interaction between the militarized, slaving confederacies of the southern interior and nonmilitarized tributary groups (Spanish-allied) ultimately led to a power differential and heightened ethnic tensions between these populations while minimizing ethnic diversity within them. Such changes would ultimately lead to Creek, Cherokee, Choctaw, and Chickasaw ethnogenesis, albeit much later in the 18th century, and are likewise responsible for the emergent "pan-Spanish Indian" ethnic identity in La Florida, invisible to us today due to the early 18th-century diaspora that culminated several decades of English-Spanish competition in the southeastern United States.

Patterns of Conflict in the Tribal Zone

In accordance with the development of a tribal zone as a consequence of state expansion, patterns of warfare, or more generally conflict, are paramount to the ethnogenetic process (Ferguson and Whitehead, 1992) and the timing, pattern, and nature of conflict in La Florida adds considerable veracity to this interpretation. The practice of warfare is anthropologically significant for a number of reasons (Ferguson, 1984; Haas, 1990; Otterbein, 1999). In colonial contexts, warfare exacerbated demographic collapse and population size stresses, forcing a broadening of mate exchange networks, while it also "crystalizes oppositions [and] separates peoples into clearly identifiable groups" (Ferguson and Whitehead, 1992: 14), which can "lead to the construction of 'adversarial identities'" (Whitehead, 1990: 360). Although the concepts of "conflict" and "tribal zone" are inextricably linked, the nature of conflict can be quite variable in different contexts and at different times within the same context. Such was also the case in Florida.

The timeline of "conflict events" in La Florida is presented in table 5.1, based on data extracted from the literature. The pattern is striking and clear. Indigenous revolts directed against the Spanish were common during the 16th and first half of the 17th centuries as populations in La Florida experienced significant social disruption. As Milanich (1999) noted, the most severe uprisings were perpetrated by first-generation converts, such as those implicated in the Juanillo Revolt of 1597 (Geiger, 1937; Lanning, 1935), and the occurrence of uprisings tracks well with those regions experiencing initial contact with the Spanish. For groups never fully tribalized (read converted), these small-scale revolts never stopped (Mayaca, Jororo).

Table 5.1. Timeline of conflict events during the Spanish mission period

Year	Event
1570	Guale revolt
1573	Guale revolt
1576	Guale revolt
1577	St. Augustine revolt
1578	Guale revolt
1580–82	Guale revolt
1584	Timucua revolt
1587	Guale revolt
1585–86	*Sir Francis Drake attacks St. Augustine*
1597	Guale revolt
1608	Five Guale caciques rebel, very minor
1608	Franciscans forced to leave Apalachee
1612	Franciscans forced to leave Apalachee
1638	Apalachee revolt
1645	Minor insurrection in Guale
1646	*Powhatan defeated by Virginians*
1647	Apalachee revolt
1655	*Robert Venables and William Penn seize Jamaica*
1656	*Timucua revolt (Potano and Utina)*
1656	*Rumored attack on St. Augustine by English*
1661	*Westo attack Santo Domingo de Talaxe and San Joseph de Sapala*
1662	*Westo attack Huyache and resettle along Savannah River*
1668	*Robert Searle raids on St. Augustine*
1670	*Charles Town founded*
1670	Guale attack Charles Town
1672	Castillo de San Marcos started
1675	Chisca incite revolt among Chacato missions
1676	Chisca raid on Ivitachuco
1677	Apalachee raid on Chisca, Chacato, and Pansacola village
1680	Westo, Uchise, Chiluque attack on St. Simons and St. Catherines Islands
1682	*French pirates raid La Chua ranch*
1683	*Grammont raid on St. Augustine*
1684	*French and English pirates attack Guale*
1684	*Scots establish Stuart's Town; Grammont reappears along coast*
1685	*La Salle's attempt at French settlement on Gulf coast*
1685	*Yamassee raid on Santa Catalina de Ahoica*

(continued)

Table 5.1—continued

1685	Two punitive expeditions sent against Creeks
1686	Marcos Delgado sent to look for La Salle
1686	Spanish attacks launched against Stuart's Town
1686	Fuentes raids Yamassee towns in South Carolina
1687	Castillo de San Marcos completed
1689	Spanish blockhouse established on Chattahoochee River
1691	Yamassee and Uchise raid on San Juan de Guacara
1694	Creeks raid San Carlos
1695	Apalachee raid against Ocmulgeee villages of Apalachicola
1695	Castillo de San Marcos completed
1702	Yamassee and Apalachicola attack Santa Fé
1702	Reprisal sent against Apalachicola
1702	Moore, with a force of Yamassee, attacks St. Augustine and destroys two Guale villages
1704	Moore attacks Apalachee and Timucua missions

Note: Italicized dates and events indicate conflicts or developments that relate to extraregional politics.

During the early years of missionization, the structural pattern of conflict changed from the internecine, almost ritualized, tribal warfare that characterized the protohistoric and presumably late precontact periods to a more localized form of armed resistance directed against the Spanish. Importantly, there is no evidence for any macroregional structure or planning; revolts were often initiated by single villages or particular leaders in response to Spanish abuses and the general deterioration of life's conditions in the wake of missionization. Often pagan elements were the primary instigators. This level of conflict with a distinctly indigenous-Spanish structure is best interpreted as a form of resistance against the incursions of the state *prior* to, but certainly a component of, complete tribalization.

Most important, after the Timucua rebellion in 1656 and concurrent with the appearance of the Westo, the defeat of the Powhatans, and the beginnings of English maneuverings south of Virginia, there is little or no evidence for any further violence directed *against* the Spanish by those communities loyal to the Crown. "Wars of resistance and rebellion" ended as part of the continued transformation of conflict within the Eastern Woodlands tribal zone. To the contrary, concurrent with the 1656 Timucua rebellion we see evidence for increasing attacks on the Spanish colony by slave raiders from the north, and subsequent reprisals launched into the hinterland were also reported (table 5.1). On several occasions, such retaliatory actions were conducted by a polyethnic, pan-Floridian indigenous contingent (Kelton, 2007: 131; Oatis, 2008: 28). Conflict had once again assumed an indigenous-indigenous structure, now pitting Spanish-allied populations against English-serving populations, a process called ethnic soldiering. That ethnic soldiering, in part an economic adaptive strategy, can hasten rapid ethnogenesis is well documented in South American Amazonian tribal populations (e.g., Ferguson and Whitehead, 1992; Hill, 1996a; Langfur, 2005; Whitehead, 1990—see also the historiography of ethnic soldiering in Parmenter, 2007) and is consistent with Wolf's (1982) perspective on the influence of European mercantilism on socioeconomic structures in emergent tribal communities (see also Ethridge, 2006).

One final comment about warfare is appropriate. It is incorrect to assume that the period prior to English expansion was completely peaceful in the Spanish colony. Violence was an ever-present condition of life on the colonial frontier. Indigenous responses to demographic collapse were commensurate with first contact, and new trading opportunities changed intergroup dynamics very early in the contact period (Waselkov, 1989). A number of raids were carried out against the interior Timucua and Apalachee, and to a lesser extent the Guale, by the Chisca, Jeaga, Pohoy, Tocobaga, Chacato, Tasquique, Apalachicola, and Amacano during the first half of the 17th century (Bushnell, 1981: 12; Hann, 1988, 1993b; Worth, 1992: 152–155, 1998b: 16–21). The Mayaca and Jororo, who lived near Lake George in central Florida, also harassed the missions and Christian Indians throughout the 17th century (Hann, 1993b). The primary distinction between the pre- and post-1650 violence was that the former was not "part of a relatively organized campaign by English settlers . . . motivated in large part by the Virginia and Carolina market for Indian slaves" (Worth, 1998b: 16). With the exception of the Chisca, whose origins are debated but certainly not local (Hann, 1988: 182–183, 402; Worth, 1998b: 52),[21] these raids were conducted by autochthonous populations with probable precontact roots in Florida that had resisted conversion to Christianity, rather than by post-mid-century ethnic amalgamations displaced from throughout the continent. This distinction may belie the ultimate stimulus for the raids and therefore imply differences of their sociopolitical effects. Although there are some references to the taking of captives (Worth, 1998b: 20), there is no indication they were taken to be sold into slavery. In fact, the capture of women and children during these short guerrilla-style assaults was commonplace during the earliest years of Spanish exploration and characterized warfare during the precontact period (see chapter 4). Many of the raids may have simply been a continuation of this long-standing practice (Bushnell, 1981: 12).

Broader changes throughout the eastern United States in response to English political machinations created shockwaves of people moving over vast distances, acting opportunistically as local situations allowed. For La Florida communities, the appearance of the Westo in 1659 marked a dramatic increase in the dimensions of their political sphere of interaction. And suddenly, English firearms and the plantation economy they sought to transfer to the Carolinas had a direct impact on the Spanish mission populations through the process of ethnic soldiering and differential firearms provisioning. As their sphere of political interaction increased dramatically, the structure of affinities and disparities also changed. Old political alliances and allegiances held little meaning as members of neighboring Christian chiefdoms who may have spoken mutually unintelligible languages seemed much more like kin than the raiding parties that descended on Florida from as far away as the Great Lakes. Such discourse is expected to shift ethnic alignments as "groups which may have been separate and even mutually hostile in one environment may be identified or identify themselves as one in a new environment *of greater heterogeneity*. The underlying mechanism is the general perceptual tendency to simplify nuances of difference, ignoring small differences and exaggerating large ones, when these begin to assume an *unmanageable degree of complexity*" (emphasis added) (Horowitz, 1975: 123). Such dialectics of difference are common features of ethnogenesis; African maroon communities perhaps provide the quintessential case study (Bilby, 1996; Horowitz, 1975; Kopytoff, 1976).

Another critical aspect of objective, etic ascription is the social distance between ascribed and ascribing communities. As the social and geographic distance between different ethnic groups increases, members of these groups are more likely to categorize others as members of a larger undifferentiated category. Similar nonspecificity in other-categorizations anchor panethnogenetic movements as well (Espiritu, 1992; Tefft, 1999), as for example, when the dominant White majority in contemporary America glosses Central and South American ethnic communities as "Hispanic" despite the great diversity contained within this category. These same psychological processes may have also been a factor in La Florida ethnogenesis. To the Westo, Yamassee, Creek, Chiluque, or Chalaque, it made little difference if they captured an Apalachee, Guale, or Timucua to sell into slavery, and certainly the English slaveholder cared even less, because the social interaction between enslaving and enslaved communities was antagonistic and one dimensional. Nuance mattered little, and nuance of interaction is what ethnic sentiments thrive on.

This dynamic certainly would have heightened the objective elements of ethnic ascription in La Florida as different indigenous communities embraced their common experiences and common historical fate, but it does not alone constitute the requisite conditions for ethnic mobilization of the scale and degree that I propose here. Ethnogenesis in this case was a matter of survival, and Spain's colonial population recognized this and acted accordingly. While the sociopolitical context was certainly conducive to the ethnic realignments I am proposing here, focus on external factors assumes no agency among the peoples directly involved in the process. To infer this level of emic discourse requires revisiting the archaeological and bioarchaeological records as well as further considering social theory that explains the process by which people come to identify with ethnic others. A focus on the emic components of ethnogenetic transformation is needed.

Six   *The Liminal Phase of Ethnogenesis*

Practice and the Lived Experience

> Ethnic affiliation may originate in an *attribution* of collective identity to them on the part of others.... But even where they have had a social identity contrived for them, subordinate groupings typically come to define their "ethnicity" as an emblem of common predicament and interest... often expressed in the reciprocal negation of the humanity of those who dominate them (emphasis in original).
> —John L. Comaroff, "Of Totemism and Ethnicity" (1987: 305)

While objectification does provide an important stimulus for burgeoning pan-ethnic sentiment, focus on etic ascription alone reduces ethnogenesis to a passive process in which the affected communities have no active role in the process. Of course, this is false. There are numerous potential responses to ascription by others and assimilative ethnogenesis is but one. In the postcolonial Americas, work by Sider (1994), Worth (2002), and Whitehead (1992) has outlined different economic or organizational strategies used by indigenous populations with vastly different outcomes. Yet all include some element related to objectification by others. At the same time, while economic factors and political discourses may ascribe ethnic communities, cultural rationalization from the ground up is absolutely essential for an identity to be reified and symbolically encoded (Cohen 1978; Comaroff, 1987), as illustrated by examples of "failed" ethnogenesis in the wake of extreme objectification (e.g., Trosper, 1981; Trottier, 1981).

Whereas the previous chapter focused exclusively on etic, objective elements that created the conditions for ethnogenetic change among Florida's Christian communities, this chapter focuses on the evidence that these communities were actively forging a shared social identity from the bottom up, an identity that was not strictly contrived for them by others. I previously noted the lack of direct historical evidence for burgeoning ethnic factionalism, competition, or ethnic boundary impermeability among Apalachee, Timucua, and Guale communities during the latter half of the 17th century. Such negative evidence should not be overinterpreted, but nonetheless it implies that an active process of emic identification may have been occurring. In other words, La Florida's communities were not only treated as monolithic by the Spanish, the English, and English-allied tribes but they also began to see themselves this way as well. In fairness, it is decidedly difficult to outline what historical evidence of ethnic factionalism would actually entail. The lack of conflict among Christian communities was noted in the previous chapter; however, this alone does not in any way guarantee that ethnic homogenization would occur. Much more personal and multiscalar forms of evidence are needed.

In the following chapter, I outline the reasons these communities may have embraced their common historical experiences and position within the Eastern Woodlands tribal zone and actively forged new bonds of community sentiment that spanned Spain's Catholic colony. More important, this chapter also outlines the evidence that this process of ethnogenetic transformation was ongoing at the time the missions were destroyed in 1706. Instrumentalist, primordialist, and practice theories of ethnicity are defined and discussed, and additional material data are presented that bolster the analyses of phenotypic variation presented in chapter 1. A greater range of bioarchaeological data is considered in light of the practice theory of ethnicity, which views experience of the world and the associated adjustments to life's expectations as the ultimate mechanism of ethnic transformation.

Primordial and Instrumental Ties

One common thread throughout the diverse and dynamic colonial and modern ethnogenesis literature is recognition that in the face of uncertainty and objective ascription, members of distinct social groups will minimize internal differences as part of the "blurring of distinctions between some groups and the sharpening of distinctions between others" (Horowitz, 1975: 127). In the wake of burgeoning and redefined interethnic conflict within the southeastern U.S. tribal zone, in concert with the stresses imposed both socially and biologically by demographic collapse, there may have been a concerted need for ethnogenesis among La Florida's indigenous communities, a need with both instrumental and primordial qualities. For example, in response to intensifying slave raiding and progressive demographic collapse, populations in La Florida may have recognized their common defensive interests and shared historical fates, and actively inculcated a broadening base of ethnic identity to bolster their notion of "community." This suggests an active, conscious manipulation of identity as part of the adaptive need for survival, although one could argue whether common defense constitutes a material interest in the instrumentalist mode of thinking. However, while such an interpretation makes intuitive sense, there is little in the historical record to support this aspect of ethnic sentiment—not surprisingly, of course—and it assumes a degree of communication and provincial-level organization for which little documentation exists. This is one of the primary critiques of instrumentalist thought: How is it that diverse groups of people come to recognize their common material interests? Emphasis on elites as brokers of ethnic identity does little to resolve this issue in La Florida. The historical record is remarkably silent about the charismatic leadership of Florida's indigenous communities. There is no sense of "ethnogenetic artifice"; in fact, elites were decidedly atavistic and traditionalist in orientation (Worth, 1995, 2002). Therefore, while the idea is theoretically possible and probable, just because a people had a reason to act collectively does not guarantee that they did so. Ethnic mobilizations often lack surface level rational coherence.

Another facet of ethnic identity theory that may pertain to the subjective component of ethnogenesis derives from the primordialist literature (Bromley, 1974; Connor, 1978; De Vos, 1975; Epstein, 1978; Geertz, 1963; Isaacs, 1974; Keyes, 1981) which seeks to explain why ethnic identity manifests, and in the particular historical tradition in which these approaches are couched, why ethnic consciousness reemerges within nations despite Marxist, class-based predictions to the contrary (Comaroff, 1987; De Vos, 1975; Epstein, 1978; Geertz, 1963; Isaacs, 1974; Roosens, 1989). In addressing issues of function, primordialists place primary emphasis on the emotional and psychological aspects of ethnic identity; the function, in this case, elicits "psychological security . . . a feeling of belonging, a certainty that one knows one's origin" (Roosens, 1989: 16), and "a feeling of continuity with the past" (De Vos, 1975: 17). Primordial sentiments are *involuntary*, coercive, and ineffable

(Jones, 2002: 65; McKay, 1982: 398), and ethnic symbols are highly emotive. As a factor in understanding emic ascription, primordial perspectives suggest a purely psychological explanation related to the "overwhelming . . . universal, human, psychological need for a sense of belongingness and self-esteem" (Jones, 2002: 66), a rarity perhaps when all you have ever known has been indelibly transformed, from the social to the ideological. In this sense, and regardless of any objectifying aspects, when faced with dysfunctional or fragmented communities such as those which characterized many former chiefdoms and villages during the latter half of the 17th century, Native Americans may have actively sought new ethnic bonds to satisfy the intense and enduring human desire for rootedness. However, these psychological aspects of ethnic sentiment are somewhat reductive and of limited analytical value. Ethnogenesis is reduced to an event that happens because of psychological need alone, with little hope of operational testing.

Despite these limitations, both instrumental and primordial perspectives position Native American communities as actively engaged in the ethnogenesis process. Instrumentalist or circumstantialist perspectives place greater emphasis on external forces, the "other-categorizations," and emphasize how shared capital interests are internalized as a call to ethnic group solidarity in pursuit of those interests. Primordial perspectives are highly internalized, abstract, and psychological in orientation, a universal "given" that merely needs description rather than justification. Consistency with theory, however, is no substitute for more direct historical evidence that ethnogenesis was occurring. In attempting to bridge the theory-data divide, an appeal to instrumentalist or primordialist perspectives does little to advance the discussion because both ultimately are rooted in motivation rather than process. Overcoming this impasse was the stimulus for Bentley's (1987) influential practice theory of ethnicity in which he attempted to explicitly link feelings of ethnic solidarity with an underlying basis to explain how ethnic sentiments develop, and not just why they develop. With my primary focus on ethnogenesis, I share a similar concern with Bentley in trying to understand how a new ethnic identity may have been embraced by indigenous populations in Spanish Florida.

The Practice Theory of Ethnicity

Bentley's (1987) practice theory of ethnicity provides a useful approach for understanding the mechanisms by which distinct ethnic communities "submit" to ethnogenesis. Building on Bourdieu's (1977: 72) concept of habitus as "systems of durable, transposable dispositions" and the use of habitus to explain how conflict arises, Bentley (1987: 27) extrapolated habitus from a mechanism of misunderstanding to one of solidarity: "Bourdieu argues that objective conditions of existence, mediated by systems of symbolic representations, generate in different persons dispositions to act in different ways. . . . Extrapolating from Bourdieu's analysis, we may hypothesize that consciousness of affinities of interest and experience embodies subliminal awareness of objective commonalities in practice." It is through these similarities of experience and resulting shared unconscious behavioral dispositions and outlooks that feelings of ethnic sentiment manifest. Bentley (1987, 1991) merely linked the structures embedded in the habitus to feelings of ethnic solidarity as "individuals, in effect, flock to others of like background, among whom their identity is not challenged by different perceptions of the world" (Stone, 2003: 38). As discussed below, this tacit assumption represents both the success and failure of a practice theory of ethnicity.

In attempting to advance the study of ethnicity, Bentley clearly refocused attention on the content of an ethnic group (see also Jenkins, 1997) where the content is in some way related to sharing a habitus. This particular aspect of his practice theory of ethnicity has been the subject of some criticism (Banks, 1996; Jones, 2002; Orser, 2004; Stone, 2003; Yelvington, 1991) because of its "whole-culture" essentializing focus on a largely unconscious basis for human action (ethnic groups form due to sharing a common habitus—see Jones, 2002, 2007; Orser, 2004), leading Jenkins (1982: 272) to quip, "The concept of the habitus has not survived the appropriation of Bourdieu by the English-speaking world. The vacuum left by its passing has been easily filled, however, by the notion of *culture*" (emphasis in original). In response to these critiques of habitus as culture, Bentley (1991) clarifies that the two are never isometric; the relationship is not unidimensional and deterministic; and that habitus itself is not a monolithic, ever-present psychological essence but is malleable, suppressible and context-specific despite being completely subconscious. Habitus is, according to Orser (2004: 132), "a set of bits and pieces, partial connections between people and things . . . [that] does not govern all human behavior . . . [and] may recede into the background in specific sociohistorical situations in which codified rules govern practice." To Bentley (1991), the distinction between habitus and culture is attributed to deep versus surface structures; the two are interrelated at a structural level but one is not completely deterministic of the other. Jones

(2002, 2007) has further elaborated this point and incorporated, although not nominally (Orser, 2004), elements of Bourdieu's concept of "field." In other words, while Bentley clearly refocused attention on the content of an ethnic group (contra earlier Barthian theory), Jones (2007) clarifies that the form of the content is historically and socially contingent (Jones, 2007: 50–51). In short, habitus does not equal ethnicity because habitus may be an integral component of many aspects of group identity in differing social domains (Yelvington, 1991).

That habitus does not constitute the totality of practice has also been criticized (Banks, 1996; Jenkins, 1992) and Bentley's apparent wholesale uncritical adoption of Bourdieu is therefore problematic to some (Yelvington, 1991; Jones, 2002, 2007). For example, Banks (1996: 47) describes habitus as "a black box—its workings are mysterious, its contents vague and it seems at times to be a repository for Bourdieu to hide an ill-defined theory of psychological motivation." Habitus has also been described as a "theoretical *deus ex machina*" (Orser, 2004: 129; see also Jenkins, 1992). Jenkins (1982, 1992), in particular, raises a number of issues with the content of Bourdieu's practice theory, not the least of which is its apparent determinism and circularity and Bourdieu's failure to explain how habitus actually produces practices. To Yelvington (1991: 168), this confusing and ephemeral relationship between habitus and ethnicity is problematic because "similarities in habitus do not guarantee ethnic sensations, and differences in habitus do not preclude identification." A similar opinion was presented by Spencer (2006: 104): "within a group, habitus may be developed differently, dependent on such factors as education, occupation, regional differences, exposure to different groups and differences in age group." This implies that there must be something more to ethnic affinity than shared dispositions which comprise the habitus in part or in total.

The problem, according to Orser (2004), is not with Bourdieu or the concept of habitus itself but with the incomplete reading of Bourdieu's work by social archaeologists and social theorists who co-opt it in ways beyond its original formulation and intent.[1] In particular, by focusing upon Bourdieu's earlier research (1977) where habitus was first outlined, those who purport to study practice in the archaeological record ignore Bourdieu's critical concepts of capital and field that were developed in Bourdieu's (1984) later writings (Jenkins, 1992; Orser, 2004). *Capital* refers to power and economic or symbolic energy that is acquired and expended by individuals, whereas *field* refers to the domains/social fields/social contexts within which capital is utilized, the "site of struggle and domination" (Orser, 2004: 136). It is the pursuit of capital, in all its forms, within a specific field of interaction that produces habitus, and habitus is the instrument that structures similar responses in different fields; it "shapes the individual's subjectivity and constrains their behavioural repertoire" (Spencer, 2006: 101) and in so doing establishes "what is possible and what is not possible . . . within a specific social position" (Orser 2004: 132).

The simplest way to explain these interrelationships is as follows: people pursue interests (capital), whether these interests are material (money) or immaterial (power, esteem), in different arenas of social interaction (fields); and they tend to do so, regardless of the particular nature of the interest or the arena in which it is being pursued, in a similar manner using a set of expectations learned in early life about the way the world works (habitus). These expectations are largely unconscious and continuously adjust to and adjust the field of interaction itself. This is pertinent to studies of ethnogenesis because people gravitate toward others with similar unspoken understandings of the world and this leads to feelings of community solidarity and ethnogenesis. This is clearly the intent of Bentley's theory: "Overlaps in the behavioral repertoire of peoples having characteristically different experiences (and habitus) are likely to give rise to invalid assumptions of mutual understanding. . . . [E]xperience of distorted communication can generate feelings of discomfort, of alienation, of hostility toward or, among the more reflective, of not knowing the other" (Bentley, 1987: 34). This to me reads as "cumulative definition through interaction" (Hudson, 1999: 13), which is remarkably consonant with Barth's (1969) influential perspective on ethnicity as well as Cohen's (1985) on the symbolic nature of community. Jenkins (2004: 42) refers to this as an "internal-external dialectic." It is only in the context of interaction with difference that one truly *thinks about* the way he or she sees the world, and this, as Jones puts it, reproduces and transforms ethnic identities. Such a view is consonant with Jenkins's (1997: 54) declaration that "the boundaries of a collective identity are . . . taken for granted until they are threatened," a sentiment Cohen (1985: 69) extrapolated to the community level: "people become aware of their culture when they stand at its boundaries: when they encounter other cultures, or when they become aware of other ways of doing things, or merely contradictions to their own culture." Given my previous emphasis throughout this book on migration and changing patterns and scales of social interaction networks, such a theoretical perspective has much to offer. It was only in the context of slave raiding and the sharp dichotomy between Spanish and English alliance that ethnic identities become catalyzed (chapter 5). Similarity and affinity are relative concepts.

If the relationship between habitus and ethnic identity is contested, and the relationship between habitus, capital, and field is difficult to operationalize, I can withdraw somewhat from Bentley's full position and focus instead on the basic elements of communalism as outlined in more general sociological theory. For example, Bentley notes that "sensations of ethnic affinity are founded on common life experiences that generate similar habitual dispositions" (Bentley, 1987: 32), which indicates that experience of the world with all its objective constraints helps create the content of an ethnic group (Banks, 1996: 46). The implications of this statement are uncharacteristically straightforward. In discussing the "notions of practice and habitus as the basic factors shaping the structure of all social phenomena," Isajiw (1992: 4) further clarifies that "ethnicity is something that is being negotiated and constructed in everyday living. Ethnicity is a process which continues to unfold . . . [and has] much to do with the exigencies of everyday survival. It is constructed in the process of feeding, clothing, sending to school and conversing with children and others." Bentley characterizes these elements of practice as the "rhythm of living," behaviors that are "inscribed, habitual, and significant, but at a level normally far removed from consciousness" (Bentley 1987: 33). It is these sorts of quotidian ways of doing that practice theory focuses on, linking the mundane, daily, subconscious rituals and practices with macroscale processes, the former a microcosm of the latter (see Lightfoot et al., 1998; Pauketat, 2001).

Despite the fact that social archaeology has moved beyond ethnicity as a focus of investigation—concentrating more on individual identity related to gender, status, and age, for example (Gowland and Knüsel, 2006)—archaeologists have clearly embraced this component of practice theory wholesale as it fits very well with the types of data they frequently work with, from simple household practices to regional communal ceremonial events, both of which are highly visible in the material record of the past (see Lightfoot et al., 1998; Pauketat, 2001, 2003; Shennan, 1993). In other words, the way you dress, carry yourself, produce technologies, and cook food, as well as the micro- and macroscale tempo of life, including landscape use and movement as well as communal ceremonial activities, are all integral components of social phenomena such as ethnic groups. What this ultimately reduces to is experience of the world, the "objective conditions," and not just in the sense that propinquity equals solidarity (see Isbell, 2000) but rather the tendency for existential *communitas* (Turner, 1969) to effect symbolic categories of belonging, to manifest an esprit de corps, with the ultimate result being the formation of a normative *communitas* (Turner, 1969), an imagined community with social structure.

Given this theoretical background, we can now return to the archaeological and historical records. Decades of research have produced an extensive record supporting the supposition of ethnogenetic transformation among La Florida's indigenous communities. These data not only indicate a more similar postcontact life experience for the Apalachee, Guale, and Timucua but also support the timing of the transition from the separation to the liminal phase. By necessity, the discussion that follows must focus on those elements of life visible in the archaeological record. It is not my intent to be exhaustive but rather to develop a general image of the social experiences of the converted, emphasizing the micro and the macro, the social, the ideological, and the practical.

Life and Experience in Spanish Colonial Florida

Space does not allow a reinterpretation of the decades of historical and archaeological research on Spanish Florida within the context of Bentley's (1987) practice theory of ethnicity. Much of this literature was published prior to or concurrently with Bentley's influential paper, and issues of ethnic identity have primarily focused on Spanish Creole community formation. Nonetheless, some broad statements can be made about the identity transforming nature of demographic collapse in the context of imperialist state expansion such that succeeding generations of Christians born into the mission environment identified commonalities with ethnic "others" to the same extent that their ancestors may have identified differences. Jerald Milanich (1999: 130) put it quite well:

> making the initial converts was the most difficult part of the process. After one generation, children born to Catholic parents at mission villages were born into the church. Catholic beliefs, rituals, and iconography replaced native ones as Franciscan friars replaced native priests. When older generations faded away, so did many native religious beliefs. . . . As a consequence, much of the old and traditional was forgotten.

It is in just such a context, when combined with intense objectification by European powers and their respective indigenous allies, in which feelings of solidarity are imbued with an ethnic sentiment. In part, the generational amnesia that comes with generational turnover may also explain the timing of the transition to the liminal stage. Individuals reaching reproductive age around 1650 would have been born and lived their entire lives within the Spanish system.

It is fair to say the lifestyle changes were fairly comprehensive and multiscalar, affecting the daily experiences as well as long-term, life course expectations of Native American populations in Florida. Part of this transformation can be directly related, at a coarse level, to simply living within the jurisdiction of an imperialist state and all of its encumbrances. For example, very early in the 16th century the following "guidelines" were offered for the proper treatment of Indians as well as their behavioral expectations:

> Proper church facilities are to be provided and the Indians must attend divine service daily. It was ordained that the day should begin in the church and the Indians were not to be required to rise before daylight. . . . Confession at least once annually (if possible) was provided for. All children were to be baptised within a week after birth.
>
> It was prescribed that the Indians work in the mines for five months and then be given a rest period of forty days. They were allowed to have their dances on Sundays and feastdays. . . . The custom of having plural wives was ordered discontinued. . . . Pregnant women were . . . allowed to nurse their children for three years. The Indians were to be supplied with hammocks and not allowed to sleep on the ground; and proper clothing was to be furnished them.
>
> Within a period of ten years men and women were to begin to wear clothes. (Hussey, 1932: 324–326)

While it is not possible to know how such proclamations affected the daily lives of specific communities, it was this strict mold of behavior and expectation into which all of the indigenous populations of the Americas would be fit. In Florida, such expectations seem to have been generally upheld by those faithful buried in the Catholic *campo santos*, regardless of what language they spoke, foods they ate, or customs they kept. And this assumption of a strict ecclesiastical existence would have further emphasized differences between indigenous Catholic populations and those living in the Georgia interior (to the north) or in the southern half of the Florida peninsula where Christianity never took hold and which served as a secondary vector of violence directed at the Spanish-allied communities.

Although the above quote referred to Caribbean populations, a similar sentiment was evident in Florida. Daily life centered around church and farming activities and community life was organized around the Christian calendar.² This is a dramatic change in the rhythm of precontact life that would have been shared by all converts *likely to have been buried in* the mission churches. Differences in settlement organization and subsistence practices, as well as degree of political centralization, would have created a much more diverse lived experience during the precontact period. And, as a testament to the successful imposition of a Christian existence, Bishop Calderón noted specifically about the Florida groups:

> As to their religion, they are not idolaters, and they embrace with devotion the mysteries of our holy faith. They attend mass with regularity at 11 o'clock on the holy days they observe, namely, Sundays, and all the festivals of Christmas, the Circumcision, Epiphany, the Purification of Our Lady, and the days of Saint Peter, Saint Paul and All Saints Day.... They do not talk in the church, and the women are separated from the men.... They subject themselves to extraordinary penances during Holy Week.... The children, both male and female, go to the church on work days, to a religious school where they are taught by a teacher whom they call the *Athequi*. (Calderón in Wenhold, 1936: 14)

Calderón was writing in 1675, during the liminal phase when ethnogenetic transformation was most active.

In addition to ideological conversion, many other aspects of life had changed in a manner that increased the homogeneity of life course experiences where little had existed before. The built landscape was transformed, with the mission *doctrina* with its distinctive rectangular church architecture, *convento*, *cocina*, and central plaza centering community life *bajo campana*. The symbolic ringing of the brass bell called the community together, and the preeminence of the *doctrina* as a landscape structuring symbol changed settlement patterning and population density. Whereas during the precontact period there was considerable difference in these parameters, population aggregation after mid-century increasingly brought people together at the missions, with the sedentary, nucleated community providing the ultimate plan and goal of the Spaniards. Prior to contact, considerable diversity had existed. The Apalachee were sedentary maize agriculturalists fully partaking in Mississippian society. Settlement hierarchies mirrored the political structure of the chiefdom, population density was high, and the center of the political-ideological landscape was the pyramidal mound structure such as those found at Lake Jackson. As one moved east into Timucua and Guale provinces the degree of sedentism decreased as did population density. Mound building was less developed and the populations were apparently more widely distributed; the burial of elites was more ephemeral and less visibly symbolic.

There is always the potential for bias in these historical reports, of course, and a visiting bishop who may have been trying to justify the Franciscan effort to the Crown is particularly subject to scrutiny. A more impartial reconstruction of the similarity of life course experiences can be inferred from the archaeological record, however, and here the data on ceramic types is most accessible and spans the Spanish provinces. As discussed in chapter 3, there is an enduring history of interpreting ceramic style within the context of ethnic identity theory because of the social symbolism often conveyed in stylistic designs and the techniques that go into this craft production. As a point of departure it is also important to note the active participation of native potters in this identity discourse. As detailed below, the 17th century witnessed a general homogenization of material culture styles across the Florida province, a fact first noted by Hale Smith, John Goggin, and Gordon Willey who proposed potential demographic implications (Weisman, 1992: 168). However, the interpretation of these temporal trends is debated. These data are discussed below by region, beginning with St. Augustine.

Stylistic Homogenization in Ceramics

Numerous scholars have contributed to the database on ceramic ware changes in St. Augustine where the frequency of nonlocal wares increased throughout the 17th century at the same time that local St. Johns ceramics associated with the eastern Timucua were gradually replaced by San Marcos ceramics associated with the Guale (Bostwick, 1976; Deagan, 1978b, 1983, 1990a; Hoffman, 1997; King, 1984; Merritt, 1983; Otto and Lewis, 1974; Piatek, 1985; see also Saunders, 1992, 2000b). At the post-1700 María de la Cruz site (SA-26-23), San Marcos wares comprised the entire aboriginal ceramic assemblage, perhaps unsurprising since the house was occupied by a Guale female and her *mestizo* husband (Deagan, 1983). To the contrary, the Convento de San Francisco witnessed a transition from predominantly St. Johns ceramics in the late 16th century, to an equal representation of St. Johns, San

Marcos, and nonlocal wares by 1650, to predominantly (91%) nonlocal wares by 1702 (Hoffman, 1993: 77). Explanations for the transformation in ceramic technologies in St. Augustine include the expansion of the *repartimiento* system and changes in the taxation and tribute economy (Deagan, 1978b, Piatek, 1985), the purchase of ceramics by the Spanish residents (Bushnell, 1981), demographic collapse among eastern Timucua populations and their replacement by refugees from Guale (Deagan, 1978a, 1990a), the emigration of other minority communities from throughout the provinces (Hoffman, 1993), and the increasing presence of native servants in St. Augustine households (Hoffman, 1997). Although the association between Guale sites along the Georgia coast and San Marcos ceramics is evident, and there is direct, incontrovertible historical evidence for the reduction of Guale and Mocama populations closer to St. Augustine (Worth, 1995), Deagan (1993) notes that the prevalence of San Marcos types in late 17th-century and 18th-century contexts exceeds that expected based on population statistics. Perhaps most intriguing is what Deagan observed about non-Guale refugee sites in the vicinity of St. Augustine during the 18th century. All sites are dominated by San Marcos wares regardless of the social identity of those inhabiting them. This finding suggests that something more than just trade or immigration patterns explains the ceramic transition.

A similar temporal trend in ceramic technologies extended to the rest of the eastern Timucua district. For example, at San Juan del Puerto the ceramic assemblage was dominated by San Marcos types while indigenous St. Johns and western Timucua Alachua types represented a small minority (see Deagan, 1978a: 115; Loucks, 1993: table 8.3; McMurray, 1973). Most interpret these data as a replacement of indigenous Timucua populations by those from Georgia (Hann, 1996a: 85; Weisman, 1992: 168). Hann (1996a: 86) notes that similar changes were evident in the few sites examined from the Freshwater and Saltwater Timucua districts.

Data from the western Timucua interior provides even more compelling evidence for regional homogenization in ceramic technology. Deagan (1972) documented a replacement of local ceramic types by paddle-stamped forms attributed to the Leon-Jefferson series at Fig Springs, which she proposed as evidence of the emigration of Creek populations from central Georgia into northern Florida during the middle of the 17th century. As noted in chapter 5, these dates correspond with the onset of Westo slave raids in the Georgia interior, and the movement of people between Florida and Georgia during this time period is widely assumed. Milanich (1978) noted that similar changes were evident in Apalachee at about the same time period, suggesting an influx of Georgia Creeks.[3] Milanich (1972), Symes and Stephens (1965), and Seaberg (1955) reported on the ceramic inventories of three sites associated with the mission period Potano in central Florida (the Richardson site, Fox Pond, and the Zetrouer site, respectively). The temporally antecedent Richardson site produced predominantly local Alachua and St. Johns types with little or no evidence for Leon-Jefferson or San Marcos wares. To the contrary, Fox Pond (thought to predate the 1656 Timucua rebellion) and Zetrouer produced a majority of either Leon-Jefferson or San Marcos types, respectively. Leon-Jefferson ceramics are associated with the Apalachee and perhaps the Yustaga (Boyd et al., 1951). Similar patterns were documented in Utina province (Johnson and Nelson, 1990; Loucks, 1993; Weisman, 1992, 1993) and provisionally in Yustaga as well (Hann, 1996a: 85). The two Utina mission period samples for which we have data (Fig Springs = San Martín de Timucua and Baptizing Springs = San Agustín de Urica or San Juan de Guacara) both demonstrated a majority of nonlocal Leon-Jefferson wares (Loucks, 1993; Weisman, 1992), which are found at near complete penetrance in 17th-century Apalachee sites (Jones, 1973; Jones et al., 1991; Loucks, 1993; Shapiro and McEwan, 1992; Shapiro and Vernon, 1992).

Weisman (1992: 166–168) suggests that the prevalence of nonlocal ceramics during the 17th century may represent the immigration of populations from central Georgia, noting that Timucua speakers may have lived as far north as the Altamaha River and therefore one may not expect to see a change in linguistic affiliation of mission inhabitants in the Timucua interior. Worth (1992: 171–182, 1998b: 36) proposes a more immediate source of an immigrant homeland, inferring a backflow of personnel from Apalachee or Yustaga eastward in response to demographic collapse. He also allows for "some form of aboriginal cultural influence from Apalachee" (Worth, 1998b: 36) rather than direct resettlement. This allays Hann's concerns about explanations that invoke large-scale population movements because there is little documentary evidence of such a grand reshuffling of people or immigration of peoples from central Georgia bearing the Lamar tradition. In either case this leaves unexplained the transition from Fort Walton to Leon-Jefferson ceramics in Apalachee province itself, which is also thought to reflect increasing influence from central Georgia during the late precontact period (Scarry, 1985; Shapiro and Vernon, 1992: 50).

Through the lens of ethnogenetic theory, some of these discrepancies are accounted for. Just because the distribution of ceramic types seems to correspond well with protohistoric period descriptions of political, ethnic, or linguistic boundaries does not then mean that changes in the distribution of these types through time *must* be anchored to these ethnic identities and their movements. Change may have been in situ to an extent. While surely some of the stylistic homogenization does reflect the direct movement of peoples or the exchange of materials due to tribute, these data could also be interpreted as a reflection of changing practices of ceramic manufacture as part of the transformation of social identities into a unified pan-Spanish Indian identity during the 17th century. Weisman (1992: 168) did allow for this explanation: "Who were the mission Indians—migrants, mixed populations of migrants and indigenous Florida Timucuans, or native peoples who across the provinces began to make and use stamped pottery with grog-tempered paste?" In other words, the temporal transformation in the material world may reflect an active process of identity manipulation such as that documented by Voss (2005) and Bell (2005) and which provided the operational model used in this monograph. Deagan (1972: 42) noted a similar opinion about the appearance of Leon-Jefferson ceramics and stamped pottery in 17th-century mission contexts, suggesting this represented "a pan-Indian response to a new set of interaction patterns."

This brief foray into material culture provides a relatively convincing argument that significant changes in Native American interaction patterns were evident during the 17th century. However, in fairness, there is no formal way to evaluate the veracity of the different explanations for the patterns observed. Trade, migration, consumption patterns, population aggregation, or perhaps a combination of each of these factors helps explain the archaeological record. I do note, however, that ethnogenetic theory predicts similar changes in the pattern of stylistic variation. As ethnogenesis commenced among communities, social solidarity was reflected in material homogenization such as that documented throughout Spanish Florida. The relatively coarse temporal framework provides poor resolution to the issue of timing, however. To address this element of the ethnogenetic model more directly I return to the bioarchaeological record, which does provide clear evidence in support of the mid-century transition to the liminal phase. As will be demonstrated below, the lived experience of Florida's Catholic populations converged post-1650 at a level reflecting an unfortunately poor quality of life. Two types of data are discussed: burial practices and health experience.

The Bioarchaeology of the Lived Experience

Burial Practices

The ceremonies surrounding death are among the most emotive and highly charged with identity symbolism. Funerary practices are often viewed as cultural compulsives and they represent an aspect of human social life that would have effected a powerful symbolic solidarity among Christianized communities throughout Spanish Florida. Although whether shared funerary practices directly cause feelings of *ethnic* solidarity *is* debatable, as an integral component of religious experience, the homogenization of mortuary practices may have, in a highly visible way, forged a stronger sense of community when shared by disparate groups over long distances. Although I am unaware of any historical text describing the funerary rites within a Floridian Franciscan church, ideological conversion visibly altered the mortuary program throughout La Florida as evident in the archaeological burial record (see Larsen, 1993, for a review of excavated mission churches with cemeteries). And in this regard, the evidence from Florida and Georgia is clear—heterogeneity yielded to newfound conformity.

The first major change in the mortuary program was the physical location where people were interred. Burial in Spanish Florida occurred either under the floor of the mission church in unmarked shallow graves such as at Santa Catalina de Guale (Larsen, 1990, 1993), Santa Catalina de Guale de Santa María (Saunders, 1993), San Pedro y San Pablo de Patale (Jones et al., 1991), San Luis de Talimali (Shapiro and Vernon, 1992), and San Martín de Timucua (Hoshower and Milanich, 1993; Saunders, 1993; Weisman, 1992, 1993) or outside the walls of the church within a *campo santo* such as at Nombre de Dios (Seaberg, 1991), San Pedro y San Pablo de Potohiriba (Jones, 1972), San Miguel de Asile (Jones, 1972), San Lorenzo de Ivitachuco (Jones and Shapiro, 1990), and San Damián de Escambé (Jones and Shapiro, 1990), some of which may have been covered with a pavilion (for example, Santa Fé de Toloca; Johnson, 1993). As many of the presumed nonchurch cemeteries were inferred based on limited test excavations, it is possible that additional research would clarify the spatial relationship between burials and the mission church (see, for example, Marrinan, 1993, for a revised perspective on mission Patale). In some cases the mortality rate was so high that the church burials extended outside its walls (Marrinan et al., 2000). Nevertheless, burial activities always occurred at the site of the mission *doctrina* in affiliation with the sacred space of the church building itself.

Burial positioning was also fairly standardized (figure 6.1). Individuals were buried in a supine position, with hands folded across their chest sometimes as if in prayer, with their feet facing the altar. Bodies were placed in rows oriented parallel with the long axis of the church and separated by a center aisle. In most cases, overcrowding due to high rates of mortality erased any signs of this original spatial structure. However, rows were visible at San Pedro y San Pablo de Patale (Jones et al., 1991; Marrinan, 1993) that were probably used for the burial of specific lineages (Stojanowski, 2005d) in accordance with the use of sepulturies in the Old World (Douglass, 1969).

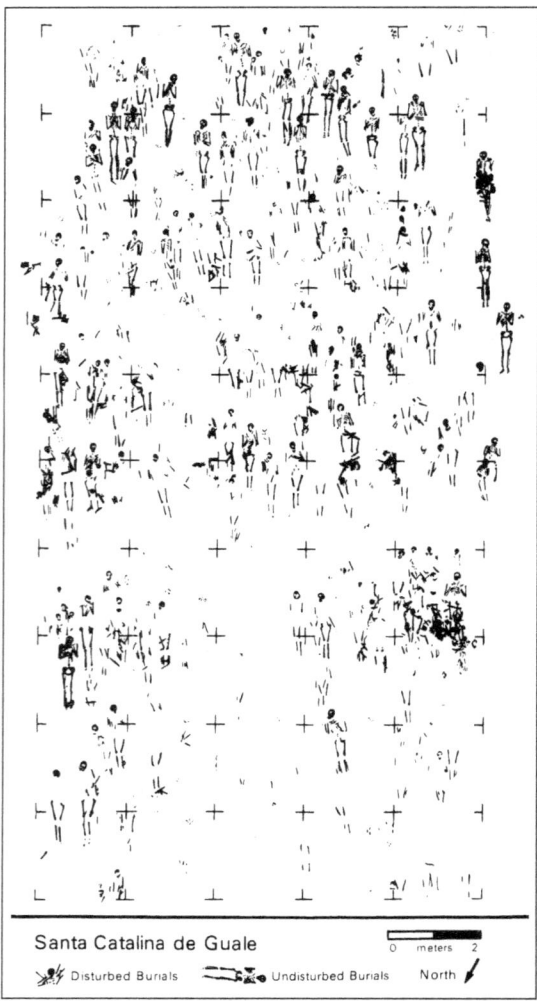

Figure 6.1. Three examples of mission-period church cemeteries demonstrating the homogeneity of spatial orientations throughout the different mission provinces. (A) Santa Catalina de Guale, (B) Santa Catalina de Guale de Santa María, (C) San Pedro y San Pablo de Patale.

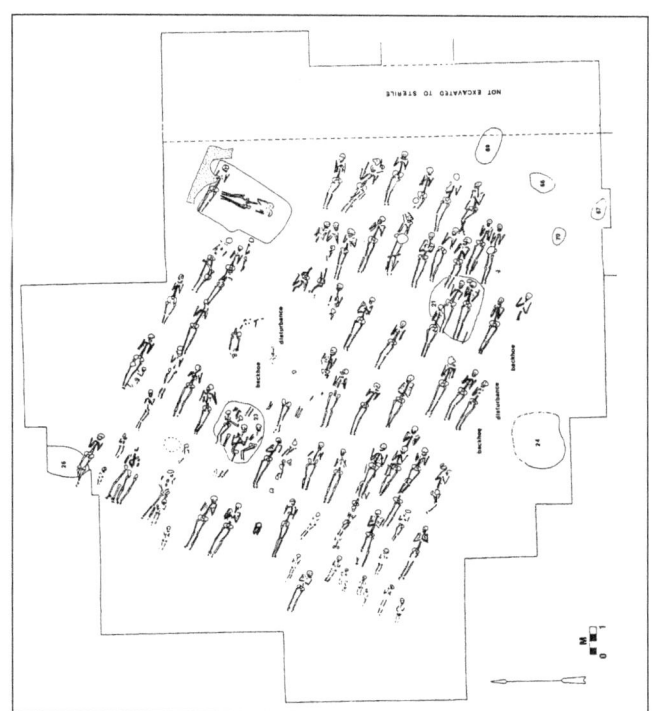

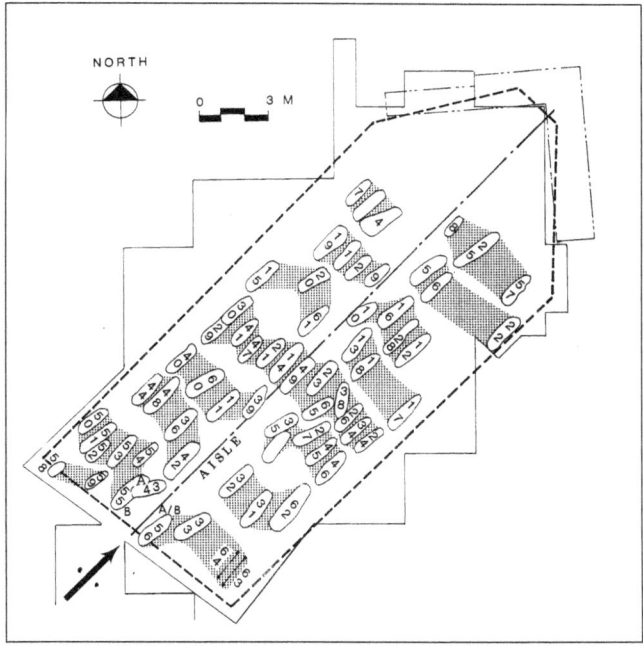

Rows can also be seen in the church of Santa Catalina de Guale de Santa María, although a center aisle is not visible (Saunders, 1993), at San Martín de Timucua (Hoshower and Milanich, 1993; Weisman, 1992, 1993) and at the O'Connell mission site (Marrinan et al., 2000). Grave goods were few, in keeping with the canons of the Christian church, and included items of personal adornment but nothing particularly ostentatious. Small European trade beads dominate the mortuary assemblages. Burial shrouds may have been used in some cases, and some high-ranking individuals were buried in coffins at places like San Luis (Larsen and Tung, 2002; Shapiro and Vernon, 1992). Prestige was recognized by placement near the altar, a New World implementation of *ad sancto*.[4] Nonetheless, the general style of burial changed very little, whether elite or commoner, Apalachee or Guale, early 17th century or late 17th century.

To appreciate the transformation (and homogenization) in burial practices one has only to contrast the above description with the variety of burial types found during the late precontact period, as represented in the data included in this study. In Apalachee province alone, the precontact samples represent homestead burials (Killearn Borrow Pit), elite pyramidal mound burials (Lake Jackson), and small mound burials (Snow Beach). Along the Georgia coast, the Irene Mound site had two distinct burial complexes, a large circular mound and a subterranean mortuary (figure 6.2). Other sites in Guale and Timucua provinces indicate the use of charnel houses and secondary burial, lineage mound burials with or without a central tomb, partially subterranean structures, single versus multiple interments, and extreme variability in grave inclusions and spatial organizations (Caldwell and McCann, 1941; Gardner, 1966; Hutchinson, 1996, 2006; Hutchinson and Mitchem, 1996; Jones, 1982, 1991; Larsen, 1982, 1990, 2002; Magoon et al., 2001; Milanich et al., 1984; Sears, 1959, 1967; Shapiro and McEwan, 1992; Thomas and Larsen, 1979; Thunen and Ashley, 1995; Wilson, 1965). Uniformity of spatial arrangement and body alignment is completely lacking in the precontact record, although the meaning of the variability is uncertain. Protohistoric descriptions of mortuary rituals are fairly common—for example, those provided by the Franciscan Ávila in 1597 (Geiger, 1937: 106; Oré, 1936: 91), Laudonnière in the 1560s (Bennett, 1975: 14–15), and Juan Ortiz the Nárvaez survivor (Elvas in Bourne, 1922: 28–29) (see also Cushing in Bushnell, 1920: 117; Hann, 1996a: 105–107, 117–119). These narratives evince an image of fairly complex and idiosyncratic rituals. The ethnographic imagery of LeMoyne/DeBry also includes a purported burial ceremony for a Timucua chief (see

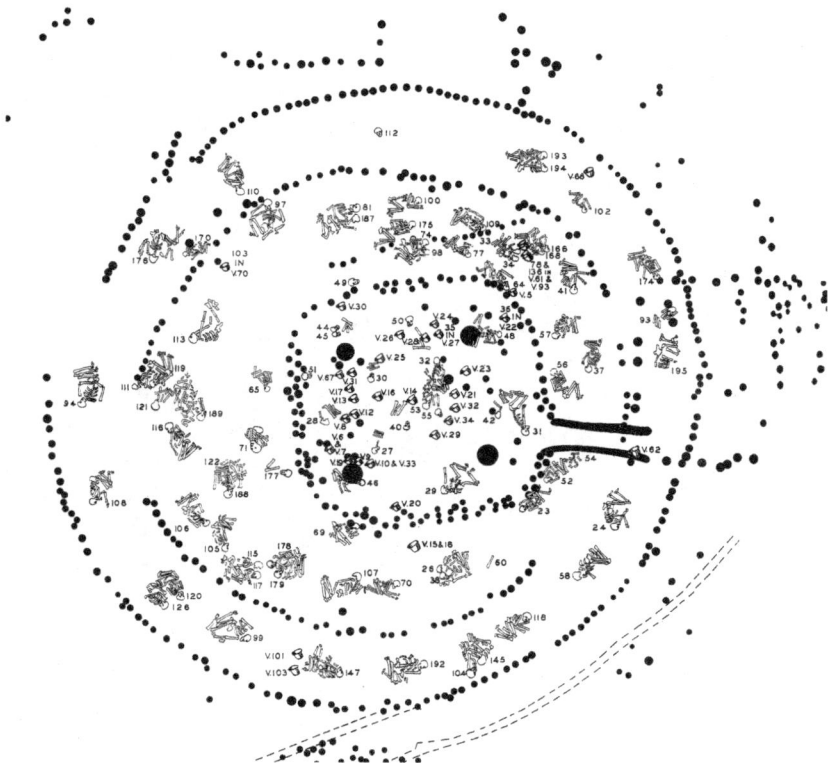

Figure 6.2. The Irene Mound mortuary complex. Figure modified after Caldwell and McCann 1941 (26).

Lorant, 1946: 115, figure 6.3), which should, of course, be interpreted with caution (see Milanich, 2005).

Health Experience

Although social archaeologists have advanced the theoretical discussion of "the body as lived experience," data on pathology and behavior derived from skeletal analysis is not typically viewed through the lens of Bentley's practice theory of ethnicity. Although somewhat crude, I believe these data do have something to contribute toward an understanding and reconstruction of the life course because health and general wellness figure prominently in an individual's outlook and experience of the world. Therefore, shared health experiences can be viewed as contributing to the development of indi-

Figure 6.3. "Burial Ceremonies for a Chief or Priest" (after Lorant 1946, 115).

vidual and community expectations and assumptions about the life course. And, as with the ceramic data discussed above, bioarchaeological signatures of health and disease have not been interpreted with respect to Bentley's practice theory of ethnicity. Nonetheless, these data on pathology and diet have been published and can be used to generate a general framework for comparing the health experiences of different ethno-linguistic communities throughout La Florida. These data have been summarized in table 6.1, as culled from various sources. At the outset it is critical to recognize the potential sampling biases reflected in these data, an unfortunate but unavoidable mitigating factor.

When these data are considered for late precontact, early mission period (pre-1650, separation phase), and late mission (post-1650, liminal phase) populations, the evidence suggests a homogenization of health experience after 1650. Prior to this time period, some groups such as the Apalachee seemed to have been spared the worst of the Columbian exchange, and therefore paleopathological signatures directly contribute to an understanding of the *timing* of ethnogenetic change among Christian communities. There are several considerations. First, data can be compared within regions across the contact transition to develop a profile of the regional experience

Table 6.1. Aggregate pathology and health data from La Florida samples

	Georgia LPC	SCDG	SCDG-AM	P	SL
Hypoplastic Defects					
% Individuals Affected[1]	86	88			
% Teeth Affected[1]					
XI1	71	71	36[5]	3	13
XI2	68	74	21[5]	3	
XC	74	78	38[5]	11	
NI1	44	43	19[5]	0	
NI2	56	52	19[5]	0	
NC	78	87	51[5]	18	61
Mean Hypoplasia Frequency[1]					
XI1	2.26	2.38			
XI2	2.22	2.09			
XC	2.18	2.29			
NI1	1.70	2.08			
NI2	2.08	1.96			
NC	2.41	2.42			
Mean Hypoplasia Width[1]					
XI1	.65	.58	.86[2]		
XI2	.58	.70	.65[2]		
XC	.63	.73	.54[2]		
NI1	.55	.41	.68[2]		
NI2	.50	.46	.72[2]		
NC	.72	.73	.78[2]		
Cribra Orbitalia %					
Total	3.1[4]	14.0[4]	22.9[4]		
Juvenile	6.1[4]	21.7[4]	73.3[4]		
Porotic Hyperostosis %					
Total	3.3[4]/6.2[2]	15.8[4]/26.5[2]	21.1[4]/27.2[2]		21[3](occ)
Juvenile	0.0[4]	20.0[4]	50.0[4]		
Periosteal Infection					
Tibia by Element %	19.8[5]/15.0[6]	15.4[5]	59.3[5]/40.3[6]	3.6	
Fibula by Element %	8.3[6]		30.9[6]	7.1	
Severity	1.35[5]	1.23[5]	1.73[5]		

(continued)

Table 6.1—continued

	Georgia LPC	SCDG	SCDG-AM	P	SL
Osteoarthritis					
Cervical Elements %	16.4[5]				18[3]
Thoracic Elements %	11.4[5]				31[3]
Lumbar Elements %	24.5[5]				11[3]
Thoracic Individual %	61[2]				40[3]
Caries					
Incisors and Canines %				0	0.7[3]
Premolars %				0	0.5[3]
Molars %				2.9	7.8[3]
Overall Teeth Affected %		8[3]	32[3]/19.6[5]	1.1	4.6[3]
Premortem Tooth Loss					
Maxilla %					25[3]
Mandible %					29[3]

[1]Hutchinson and Larsen (2001).
[2]Larsen et al. (1992).
[3]Larsen and Tung (2002).
[4]Schultz et al. (2001).
[5]Larsen et al. (2001).
[6]Larsen and Harn (1994).

of the missionization process. Second, data can be directly compared among indigenous communities for the same time period, that is, early mission period (pre-1650) Apalachee versus Guale and late mission period (post-1650) Apalachee versus Guale. Data from Timucua province is less well reported and several postcontact samples such as Pine Harbor, the ossuary at Santa Catalina de Guale de Santa María, and Santa María de los Yamassee are of questionable temporal or ethnic affiliation. Therefore, the following discussion focuses on precontact Apalachee and Guale samples, the early mission period samples from San Pedro y San Pablo de Patale and Santa Catalina de Guale, and two late mission period samples from San Luis de Talimali and Santa Catalina de Guale de Santa María.

In my previous book I discussed the evidence for differential response to missionization in Apalachee and Guale provinces. Beginning with the former, there is little reason to accept the notion that European contact led to widespread and immediate declines in health among indigenous communities (Stojanowski, 2005a: 115–116). For example, despite population size estimates for the precontact and early contact period Apalachee that suggest a 50 percent reduction (Hann, 1988: 162–164), there is little biological or archaeological evidence for significant increases in mortality during the early mission period. Several observations support this. First, there is limited historical evidence for epidemics affecting the Apalachee prior to 1650 (see Hann, 1986b: 392, 1988: 175),[5] notwithstanding the population size enumerations that suggest a 50 percent reduction. Second, the Patale cemetery, the best-known early mission period church from the province, was well organized and exhibited neatly arranged rows and a center aisle (Jones et al., 1991). There were also few commingled burials. Both observations indicate that the rate of burial during the period 1633–1650 was not of a catastrophic proportion. The rate of burial as a percentage of the estimated population living within the mission's jurisdiction was the lowest for any mission church for any time period (Stojanowski, 2005a: 117). Third, phenotypic variability did not decrease across the contact period transition, which contradicts the expectations of demographic collapse and declining population size (Stojanowski, 2005d: 115–118). And finally, pathological data for Patale do not demonstrate an increase in comparison to precontact Apalachee samples (Jones et al., 1991; Storey, 1986). In fact, Storey (1986) found a decline in the frequency of hypoplasias, markers of infection, and signatures of anemia in the Patale sample in comparison to the late precontact sample from Lake Jackson. Monahan Driscoll and Larsen (1994) reported raw data by burial for Patale from which the sample frequencies in table 6.1 were derived. Patale demonstrated a very low frequency of linear enamel hypoplasia (figure 6.4), tibial and fibular periosteal infection, and dental caries.

To the contrary, for Georgia coastal populations buried at Santa Catalina de Guale, the transition to the early contact period witnessed a myriad of signatures of increased morbidity and mortality. For example, there was a marked increase in population genetic variability (Stojanowski, 2005a) that reflects immediate post-missionization population aggregation along the Georgia coast in the wake of early 16th-century epidemics. This interpretation is supported by population size data indicating a much-reduced Guale population size by the early 17th century (Lanning, 1935: 18; Worth, 1995: 13); historical accounts of epidemics directly affecting the coast (Bushnell, 1981: 13; Deagan, 1978a: 94; Dobyns, 1983; Hann, 1988: 175, 1996a: 174; Swanton,

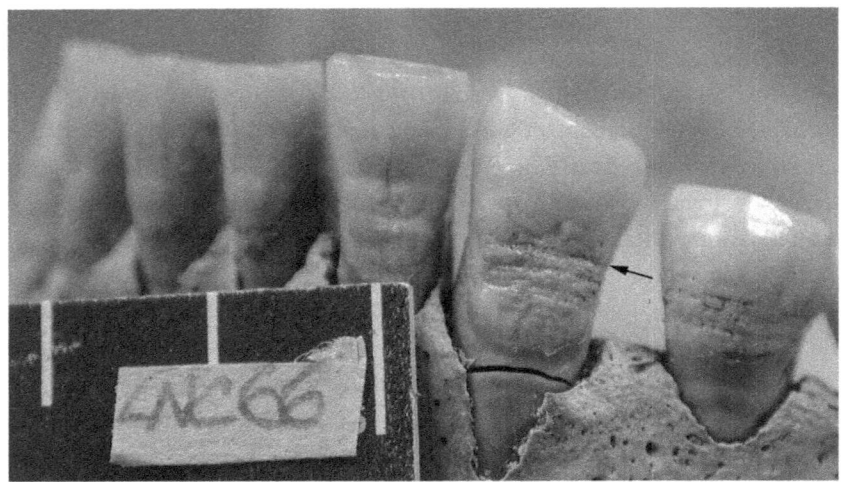

Figure 6.4. Example of a hypoplastic defect from a Florida Native American. Black arrow points to the defect on the canine. Photo courtesy of Glen H. Doran.

1922: 337); mortuary data on the frequency of commingled, secondary, and intrusive burials at Santa Catalina de Guale (Larsen, 1990), which contrast sharply with mission Patale; and bioarchaeological data indicating that community stress increased while overall health decreased at Santa Catalina de Guale in comparison to precontact period populations (Larsen, 2001, and references therein). Specifically, the total frequency of cribra orbitalia and porotic hyperstosis increased sequentially as did the severity of the infection (Larsen and Sering, 2000: 126; Larsen et al., 1992: 33; 2002: 424), while hypoplasia metrics demonstrated a more mixed signature (Larsen et al., 2002: 416). However, several indicators such as tibial periosteal reaction frequency and severity did not increase during the early mission period (Larsen et al., 2002: 425).

Regional differences in health experience during the early mission period are stark. For example, at the Patale mission the frequency of linear enamel hypoplasia for the maxillary incisors was 3 percent and the frequency of linear enamel hypoplasia for the maxillary canines was 11 percent. Comparable data from Santa Catalina de Guale cemetery indicates frequencies of 71 percent and 78 percent, respectively (Hutchinson and Larsen, 2001). In contrast to Santa Catalina de Guale where nearly half the mandibular incisors displayed hypoplasias, none at Patale did. The Patale sample also had lower frequencies of tibial and fibular periosteal infection and there was no evidence for porotic hyperostosis or cribra orbitalia. Such stark contrasts are difficult to explain by sampling bias alone (given that both samples were extremely poorly preserved) and likely reflect actual phenomena. Although the osteological paradox must be considered, it seems reasonable to conclude that the health experiences of ethnic Apalachee and Guale differed considerably during the first half of the 17th century, regardless of whether skeletal signatures of stress indicate robust or poor health (see Wood et al., 1992).

This changed after 1650, however. After mid-century, all of La Florida shared the burdens of the *repartimiento* system;[6] suffered periodic epidemics and increasing morbidity; and felt the loss of family and extended kin due to slave raiding, labor abuse, epidemic disease, and even fugitivism. Quality of life was uniformly poor, and this is supported by the bioarchaeological record where data from San Luis de Talimali suggests a regional homogenization of morbidity experiences had occurred. The cemetery at San Luis was overcrowded (as was Santa Catalina de Guale) and the best measure of morbidity experience (linear enamel hypoplasia) indicates a dramatic increase through time (61% of mandibular canines were affected—Larsen and Tung, 2002) with frequencies now approaching those documented in Guale province (Santa Catalina de Guale = 87% of mandibular canines and Santa Catalina de Guale de Santa María = 51%). Converts at San Luis also exhibited similar rates of porotic hyperostosis (~21%) and slightly lower rates of thoracic osteoarthritis (40% of individuals) when compared to late mission period Guale at Santa Catalina de Guale de Santa María (21% and 61%, respectively). Even allowing for significant sampling bias, the differences in health experience during the early 17th century are about as dramatic as one could expect. Yet these experiences converged toward a tragically high level of stress as the Apalachee were further incorporated into the fold of the mission system.

In addition to health experience, dietary homogenization also characterized the latter half of the 17th century. In particular, maize consumption intensified throughout La Florida (Larsen et al., 2002) where it had been of variable importance in prehistory (Deagan, 1978b; Jones, 1978; Larson, 1969, 1980; McEwan, 2000b; Milanich, 1978, 2000, 2004; Saunders, 2000b; Worth, 2004). Simultaneously, reliance on marine resources declined, leading Larsen et al. to conclude that "the pattern of change that we document via stable isotope analysis is one of a shift from regional diversity to uniformity and convergence of diet across the landscape" (Larsen et al., 2002: 75). And like the cultural compulsive that is the funerary ritual, foodways and preparation techniques (cuisine) may have evolved in concert, both of which are charged with identity symbolism. Therefore, from a bioarchaeological perspective, the two variables with the greatest visibility in the archeological record (mortuary practices and health experiences) both suggest a more common life course for far-flung communities of individuals during the latter half of the 17th century in La Florida.

When considered in isolation, these bioarchaeological indicators of the lived experience may not evoke feelings of ethnic unity and solidarity. Nonetheless, it is clear that the populations of La Florida did experience similar patterns of morbidity, eat more similar types of foods, and practice similar types of burial rituals as demanded by the Catholic church. At the same time, the transformation in the material realm would have provided a highly visual and effective symbol of social identity. Above, I have focused on ceramic technology because it correlates with the general picture of dietary homogenization as represented in the stable isotope data, and pottery comprises a considerable portion of the archaeologically recovered material assemblages. However, other visible changes in the material world were surely commensurate, including such highly symbolic items as dress and adornment, the built landscape, tools, and manufacturing techniques as but a few examples. When considered in concert with other sweeping changes in Native American lifestyle throughout the 17th century, not the least of which was ideological conversion, the conditions for an emic genesis are undeniable.

It must be stressed that all of these changes in the material world occurred in the context of a divided system, the two "Republics" (Bushnell, 2006), which prohibited social mobility, and prior to the development of the *sistema de castas,* which so pervades the later colonial Hispanic experience (Deagan, 1997). These changes would have been most noticeable in the earliest years of missionization when considerable variability between young and old, traditional and modern, still existed. However, as demographic collapse took its toll on local population structure, migration increased concurrently. And this, combined with the dramatic restructuring of central Florida by Governor Rebolledo in the aftermath of the Timucua rebellion of 1656 (Hann, 1996a; Worth, 1998b), would have increased contact and interaction among myriad previously distinct ethnic groups, with St. Augustine perhaps serving as an epicenter of information exchange. And it is in these invisible daily interactions between strangers that the process of identity transformation would have commenced. Phenotypically similar, sharing the burdens of the Spanish labor system with adopted Spanish surnames and perhaps attire, with histories of hardship, epidemic-related mortality of family members, life as a past fugitive or family members who were fugitives, the daily and annual rituals of the Catholic church, foodways and preparation practices—all of these in a very stark way would have united the populations of La Florida. As Milanich (1999: 144) noted, "through the missions, the Indians embraced a new faith, iron tools, the Spanish language, monogamy, and peaches." The juxtaposition of the macro and the micro in this statement, converting to Catholicism versus eating peaches, in a subtle way perfectly encapsulates the nuance of the practice theory of ethnicity.

The Spanish missions were ultimately destroyed at the hands of the English and their Creek allies during the first decade of the 18th century. The populations this book has focused on were all but destroyed. Many were killed or enslaved during the raids; some fled west, some east; and many simply disappeared from the eyes of the European observers and from their (our) history. However, this focus on ethnogenetic transformation during the 17th century does not end with the destruction of the missions because the people themselves did not go "extinct" as would a species in the biological mode of thinking. The process of ethnogenesis simply continued in its then current form, liminality, and would continue in that mode for several decades before entering a stage of separation and then reintegration. Finally, an ethnonym was created for and eventually inculcated by these peoples that we now call the Seminole. Indeed, realization that ethnogenesis was occurring among Catholic communities in Florida requires careful reconsideration of the 18th-century history of the indigenous peoples of the southeastern United States. This topic is treated in greater detail in the next chapter where I explore the reasons that the 17th- and 18th-century populations of Florida and Georgia were not as distinct as textbook colonial histories and ethnonymy might suggest.

Seven *Bridging Histories*

Seminole Ethnogenesis Reconsidered

> Although the Seminoles and Florida have been linked for many years, the tribe was part of the Creek Confederacy in Alabama and Georgia for a far longer time and was a relatively late arrival on the peninsula: the Apalachees, Calusas, Timucuans, and other smaller tribes arrived much earlier.
> —James W. Covington, *The Seminoles of Florida* (1993: 3)

> The [Seminole] tribe is an entirely post-European phenomenon, a replacement by Creek settlers of the Florida aborigines whom they eliminated in frontier military campaigns growing out of antagonisms between European powers.
> —William C. Sturtevant, "Creek into Seminole" (1971: 92)

There are books on the Creeks, the Seminoles—both at the same time—and on the Florida missions. Casual perusal of these titles may suggest that the historical period in the Southeast is compartmentalized. There is the Spanish mission period, and then there is the post-Spanish mission period when the Apalachee, Guale, and Timucua cease to "exist" in name and instead we speak of Creeks and Seminoles. While the tendency to create distinct literatures for the pre- and post-Spanish mission populations may reflect the sheer depth of knowledge and volume of historical materials available about these peoples, the tendency to divide history in this manner is also symbolic of the perceived finality of Moore's raids on the Spanish missions. In short, the narrative history is provided an easy marker, a terminus reified through the vagaries of colonial ethnonymy. Moore's destruction of the missions culminated an ethnocide perhaps, but only through the lens of European colonial historians. As Wickman (1999) argues, however, there was nothing that occurred during the 17th and 18th centuries in the Southeast that could be classified as a true ethnocide—the killing of a culture. The reality is far more complex and directly tied to the biases of historical observers and the vagaries of nomenclature used to refer to the tribes whose actions are recorded piecemeal in the historical annals.

In this sense, a neologism may be in order—nomocide, the extirpation of an identity through historical invisibility. Perhaps it is less insidious if its cause is observer bias or historical inaccuracy, but the effects of nomocide on the parsing of cultural patrimonies are equally as deleterious as complete extirpation to modern descendant communities who are systematically denied the rights afforded "aboriginals," as this chapter's opening epigraphs indicate. As historian Patricia Wickman noted, "the act of naming is also the act of creation and the 'right' to create is inherently the assumption of a profound power" (1999: 183). Indeed, Jonathan Hill's views on ethnogenesis directly address this issue, noting that it is "a concept encompassing peoples' simultaneously cultural and political struggles to create enduring identities in general contexts of radical change and discontinuity" (Hill, 1996b: 1).

However, even a casual perusal of Spanish mission, Creek, and Seminole historical literatures indicates that the colonial period narrative is not so cleanly divided. The textbook history is widely regarded as incomplete at best, if not largely ethnocentric. Indeed, I would argue, as have others (Sattler, 1987, 1996; Wickman, 1999), that the destruction of the Spanish missions by English-serving slave raiders did not end the process of social identity transformation. It is true that many residents of La Florida participating in the process of ethnogenesis were killed during the assaults on the province, enslaved and sold in the Carolina colony, or escaped these more deleterious consequences of Spanish allegiance only to reappear intermittently in the later colonial histories of French Louisiana or Spanish St. Augustine. Regardless of the ultimate fate of these individuals, however, the active process of identity transformation among La Florida's Christian communities was preempted prior to the stage of reintegration and designation of a common ethnonym. It is this latter fact that creates historical invisibility.

However, the biohistorical narrative developed in this book does not end with Moore's raids. The historical processes of social adaptation during the

Spanish period helped define the broader trajectories of change throughout eastern North America. The missions were a small but significant aspect of the saga of collapse and regeneration experienced throughout the continent, as initiated by epidemic disease, new modes of economic practice, and the impositions of various colonial entities, each with distinct motivations. But these are exogenous forces that divorce Native American history from the Native Americans themselves; it absolves their agency and active participation in the historical process. In addition, one simply cannot understand what unfolded during the 18th century without some appreciation for the patterns of biosocial adaptation of the preceding centuries. And this becomes increasingly evident when the poorly documented formation of the Creeks during the 17th century is considered (Hahn, 2004; Hann, 1996b; Knight, 1994; Worth, 2002).

In this chapter I propose that the identity transformation that was occurring among Spanish Catholic communities in Florida (Apalachee, Guale, Timucua) transcends the existence of the protohistoric ethnonyms that fill 17th-century historical narratives. The Spanish period, itself a continuation of protohistoric discourses evoking a Mississippian cosmogony, informs our understanding of the next major stage of Florida indigenous history: Seminole ethnogenesis. I specifically propose that the *ultimate* reason for Seminole ethnogenesis—that is, divergence from the Lower Creeks—is the emigration of disparate communities from Georgia into the vacant lands of the old Spanish missions. This is not a novel observation or even a deniable truth. Geographic distance then led to social distance, and this ultimately is the cause of Seminole ethnogenesis despite more proximate contributing factors which will be detailed below. Exogenous forces were certainly a significant impetus for these migrations; however, one of the primary motivations for *some* of the proto-Seminole communities to move back to Florida was far more agent-based. They were Christians and pro-Spanish (Hahn, 2004; Sattler, 1987, 1996) and they actively resettled locations in Florida where the largest populations of Spanish Florida Indians resided. The ethnic affinities of these individuals indicate that the Apalachee, and to some extent the Yamassee and Timucua, served a critical role in the migration rationale. Simply put, the beginnings of Seminole ethnogenesis can be traced to an active return to a recently vacated homeland, but not by a homogenous Apalachee and Yamassee ethnic conglomeration but by bands and families whose shared identity was in the process of becoming during the 17th century and revealed by the evolutionary analyses of gene flow and migration presented in chapter 1.

In this sense, Seminole emergence represents another stage in the process of colonial ethnogenetic transformation, coincident with a burgeoning flexibility of identity categories and the willingness to actively engage the colonial frontier opportunistically. Indeed, the conflating of a "people" with a "nation" is patently false and suggests the imposition of a European worldview where it clearly does not apply. Identity and residence were distinct for Native Americans. They were both highly malleable but independent to the point that the two were never isomorphic. Such an interpretation of indigenous community organization is bolstered by the lack of biological diversity among these populations during the precontact period (chapter 1).

This interpretation of the Seminole past is not new. Numerous scholars have noted the presence of Christian Indians in the early Seminole communities and considered their potential relevance (Craig and Peebles, 1974: 89; Fairbanks, 1978: 165–166; Hahn, 2004: 58, 77, 78, 92, 112; Sturtevant, 1971: 102; Swanton, 1922: 403–408; Weisman, 1989: 173; 2000: 303). However, "input" was generally not meant to imply "impact," and the presence, if any, of Florida's 17th-century tribes is generally seen as minor in the history of the Seminole peoples. The exception is the work of historian Richard Sattler (1987, 1996), who places a much greater emphasis on Spanish period indigenous populations in parsing the history of the Florida Seminole. Patricia Wickman (1999), working independently, has come to the exact same conclusion, and both historians hypothesized what this book has demonstrated about the biological connections between the Apalachee, Timucua, and Guale populations and the earliest Florida Seminole.

This chapter draws the various strands of early Seminole history together and interprets them within the context of the ethnic identity changes occurring among Christian populations during the 17th century. Because the database is limited and currently lacks any "Seminole" samples, this chapter is more exploratory. The data are pushed beyond the norm of interpretation and the content of this chapter should be viewed as a series of hypotheses subject to additional testing. To begin, a "straw man" history is constructed that represents a "textbook" history, one that is simplified and filtered through a Euro-American historical framework. For the nonregionalist, this should serve as a useful primer of early Seminole history. For the specialist, it may appear that things are oversimplified as an appeal to iconoclasm.

This is not the intent, as my specific area of focus (the very beginnings of the Seminole) is so very minor in the grand saga of Seminole history. The interpretation does not change the reality of later Seminole history, but it does position the Seminole firmly within a continuous biohistorical narrative linking them with Florida's precontact indigenous populations.

Seminole Ethnogenesis: A Textbook History

That entire books have been written about the Seminole means that a more simple distillation of the critical details is needed. It is not my goal to rehash these excellent summaries here. And, in fact, attempting to link the precontact tribes of Florida and Georgia with the postcontact Seminole makes the latter phases of Seminole ethnogenesis appear somewhat tangential to the development of this biohistorical narrative. As such, I focus primarily on the earliest phases of Seminole history, the period from about 1715 to 1835, when the ethnogenetic process continued in a state of liminality and which culminates in the separation of Lower Creek and Seminole identities.

Seminole history begins with the Lower Creeks, a group of loosely allied villages that evolved partly in situ from the remnants of chiefdoms in the interior of Georgia. These peoples were displaced by de Soto and other Spanish entradas and, while primarily indigenous to Georgia, also incorporated many foreign ethnic elements throughout the 17th century, including the politically dominant Muskogees who migrated from the west (see Hahn, 2004, 2006).[1] After several failed attempts at Spanish missions along the banks of the Apalachicola River (Hahn, 2004; Hann, 1988, 1996b), the Lower Creeks abandoned their homes and resettled at key locations near the Ocmulgee River during the 1680s. The reason for this migration is well known. Spanish reprisals against the Creeks for trading with the English were heavy handed, and Spanish and Apalachee forces destroyed several Creek towns during the 1680s.[2] A fort was built near the Creek town of Coweta in 1689, which heightened tensions. Throughout the next two decades conflict would continue to escalate between the Georgia Creeks and the Spanish, an escalation that culminated with James Moore's raids on the Spanish missions. By moving closer to Charles Town, Creek trade with the English was facilitated—opportunities for greater wealth abounded.

However, relationships with the English soon soured due to continued White encroachment on Creek lands, the degradation of hunting lands, epidemic mortality, alcohol abuse, burgeoning debt loads, and the increasing fear the Creeks themselves would be enslaved as a means to pay these debts (Hahn, 2004; Oatis, 2008). A shift in geopolitical alignments that now favored the Cherokee and Choctaw heightened tensions (Fairbanks, 1978; Kelton, 2007: 204),[3] and in 1715 violence erupted at Pocotaligo by a group of Yamassee, Apalachee, Yuchi, and Shawnee (Cline, 1974: 64; Crane, 1956: 170; Swanton, 1922: 98). Brims, the Lower Creek leader from Coweta, along with a polyethnic conglomerate attempted to rid the region of the English, and conflict quickly expanded throughout the Carolina frontier. The insurrection was not a success, and in the end the Yamassee War was a disastrous defeat for the Creeks and their allies, in part because the Cherokee refused to join them (Oatis, 2008). In the aftermath, many, but not all, Lower Creek towns moved back to the Chattahoochee River region and brought with them remnants of the Yamassee, Yuchi, Apalachee, and Savannah peoples.

Shortly after the return of the Lower Creeks to the Chattahoochee, individual villages and communities began migrating to the largely vacant territory of Spanish Florida. The dates are uncertain and subject to debate but appear to postdate the Yamassee War by at least several years. Continued conflict between England and Spain, French incursions, and the secession of the United States from England created a dynamic sociopolitical context in the Southeast in which allegiances were continuously strained and restructured. The 18th century witnessed a gradual distance developing between the Creeks and proto-Seminoles, defined both socially and spatially. After transfer of Florida to England in 1763, British strategies for Indian interaction were implemented—divide and rule was the norm. Preferential trading relationships, the machinations of specific village leaders, and the negotiation of treaties tended to further highlight the sociopolitical differences between the Lower Creeks and the proto-Seminoles toward the middle of the 18th century (table 7.1). And it was around this time that the appellation "Seminole" also began to be used to describe the Florida "Creeks," initially applied only to the Alachua communities but later expanded to include the whole of the Florida populace (Sturtevant, 1971: 105).[4] The correlation between the transfer of power in St. Augustine and a shift in colonial ethnonymy is no coincidence, as the English had "no cultural memory" of the groups the Spanish had spent so much time proselytizing (Wickman, 1999: 216).

The mid-18th century was a period of sociopolitical transition, with the leadership of the Lower Creeks at times representing the proto-Seminoles in negotiations with the Europeans, and other times not.[5] The Florida groups

Table 7.1. Significant events in the process of Seminole ethnogenesis

Event	Effect
Rise of Alexander McGillivray ca. 1784	Nontraditional leadership; divided Creeks internally; caused Seminole withdrawal from Creek politics
Treaty of New York 1790	Ceded Lower Creek lands; Seminole complied, driving rift between Florida and Georgia groups
Influence of William Bowles	Interrupted trading relationships; worked against McGillivray to foster anti-Americanism among Lower Creeks and Seminole
Death of McGillivray 1793	Usurped traditional leadership structure; death left power vacuum allowing greater autonomy between individual villages
Return of Bowles 1799	Further disrupted trading relationships
Tecumseh/Red Stick Nativism 1811	Anti-White sentiment spread through Upper Creek villages, Flint River Seminole, Apalachee Seminole; Lower Creeks eschewed Red Stick ideology
Creek War 1814	Americans and Lower Creeks defeated Red Stick Upper Creeks; Red Sticks fled to sympathetic Seminole; Seminole population increased greatly
Arbuthnot's Trading 1817	Fomented suspicions of Seminole, increasing attacks on Americans
Andrew Jackson Invades Florida 1818	Several Seminole villages destroyed; Seminole pushed further south, increasing contact among disparate Seminole groups; Lower Creeks aided Jackson
Florida ceded to United States 1821	Intensified Lower Creek attacks against Seminole
Treaty of Moultrie Creek 1823	Seminole agreed to move to central Florida; growing unification among Seminole; designated single speaker for negotiation
Treaty of Paynes Landing	Seminole agreed to move to Oklahoma

remained generally aloof, and it was clear that there was declining interaction between the Lower Creeks and proto-Seminoles during this time. Personality and charisma also played a role in this phase of Seminole history through the influence of such individuals as Alexander McGillivray and William Augustus Bowles.[6]

Around the turn of the 19th century the Creeks began recognizing the Florida groups as an entity distinct from themselves (Sattler, 1987: 64). This change resulted from the dissolution of unity effected by Alexander McGillivray as well as further disagreements over the negotiation of land rights and debt settlements (Sattler, 1987; Sturtevant, 1971: 105). Further fissioning pressures resulted from the American presence in the Southeast. White population expansions after the American Revolution imposed on Upper Creek towns, which changed the interaction dynamic. The Shawnee prophet Tecumseh, among others, instilled an element of nativism, anti-Americanism, and traditionalism in Upper Creek villages, which fomented into open war between the Creeks and the Americans (the Creek War, 1813-1814). An American victory in this conflict forced many of the conservative elements (the Red Stick Creeks) to abandon their Upper Creek towns and settle in Florida among the generally sympathetic Seminole. As such, the Creek War dramatically changed the composition of the Seminole villages which had prior to this time been composed primarily of immigrants from the Lower Creeks. The Lower Creeks themselves remained neutral during this conflict, which heightened the division between Lower and Upper towns. And when individuals from the latter were incorporated into Seminole communities, the antagonism transferred. The depth of the wedge driven between the Lower Creeks and the Seminoles was demonstrated in the First Seminole War (1817-1818) during which Lower Creeks actually participated alongside Andrew Jackson in assaults on numerous Seminole towns. One effect of Jackson's aggression was the migration of many Seminole villages farther south where a sense of Seminole social and ethnic community could develop in nearly complete isolation from their Creek ancestors.

In 1821, America seized Florida from Spain and it was at this time that the disparate Seminole elements in Florida formed a unified political council to negotiate with the U.S. government. Although the power of this council was initially limited, the removal of all Seminole communities to central Florida and the election of a governing body were complete by around 1835. It was during this time that the Second Seminole War erupted. Weisman (2000) proposed that a unified Seminole identity coalesced in this period—in other words, this was the time when Seminole ethnogenesis entered a state of advanced liminality in which new sociopolitical connections were established.[7] According to Sturtevant (1971: 110), by 1842, "Seminole culture in Florida evolved entirely independently [from the Creeks]," reflecting the implementation of the policies of Indian removal by the U.S. government. It was in this period that Seminole identity was actively coalescing; the stage of reintegration had begun.

The later history of the Seminole is dominated by the souring of their relationship with the U.S. government and the government's repeated attempts at removal, details of which will not be treated here. This is not to deny the importance of Seminole-U.S. relations, the influence of the Black Seminole on identity issues, or the importance of 19th-century migration histories for parsing modern Seminole tribal identities. Rather, attempts to link the Spanish and Seminole periods of history in a continuous biosocial narrative, as well as efforts to understand the processes responsible for Seminole ethnogenesis, their divergence from the Creeks, defined both etically (when the Creeks recognized them as different) and emically (when the Seminole recognized their unity), effectively sets the temporal dimensions of consideration. Multiple researchers recognize this as the colonization phase of Seminole ethnogenesis (Craig and Peebles, 1974; Fairbanks, 1978; Sattler, 1987, 1996), and understanding the motivations behind the initial [re]colonization of Florida is the primary focus of this chapter.

Seminole Ethnogenesis: Ultimate Causation

In avoiding the later history of the Seminole, the three wars they fought with the United States, the story of Osceola, the input of the African maroons during and after the American Revolution, the retreat to the south Florida swamps, and the later political history including removal to Oklahoma and the politics of federal recognition, I am focusing on a simplistic yet paramount fact: there would probably not be a Seminole or Mikasuki identity today if dissident factions of Lower Creeks had not migrated from Georgia to Florida during the early 18th century. It was this geographical distance that created the distinct sociopolitical contexts in which these communities lived. As the needs and desires of the Florida and Georgia groups diverged, social distance increased, thus leading to Creek-Seminole divergence. Of course, as the preceding paragraphs outlined, there was much more to this than simple distance and isolation (although certainly isolation policies continued to play a role). However, in seeking an ultimate cause for Seminole ethnogenesis one must focus the discussion on the motivation and rationale for the proto-Seminole emigrations, while at the same time admitting that multiple causes were likely operative.

Typical explanations for the move to Florida emphasize both push and pull mechanisms. The Yamassee War stressed the relationship between the Lower Creeks and the English, and in practice the trade in Native American slaves which had defined English-Creek interaction for the last few decades had all but collapsed by 1735 (Hahn, 2004; Kelton, 2007; Oatis, 2008). Therefore, when Lower Creeks migrated back to the Chattahoochee there was less incentive for them to stay there and maintain the status quo. A shift toward importing African slaves and away from exporting indigenous slaves revolutionized the colonial economy, and Native Americans were not benefited by this. In addition, Cherokee raids on the Creeks continued throughout the 1720s and the founding of the Georgia colony by Oglethorpe in 1732 further increased the pressure exerted by White settlers on the Creeks (Craig and Peebles, 1974; Fairbanks, 1978: 165; Sturtevant, 1971: 101). Another push mechanism was the political instability within the Lower Creeks after the Yamassee War. The Creek leader Brims assumed a more neutral position with regard to the colonial entities in the southeastern United States and adopted a strategy that involved opportunistic and shifting allegiances.[8] Others within the Lower Creeks were decidedly pro-Spanish or pro-English in allegiance (Hahn, 2004; Sattler, 1996: 43; Sturtevant, 1971: 101).

At the same time, the Spanish invited the Lower Creeks to resettle the old mission fields to create a physical buffer for St. Augustine. Spain sent emissaries to the Lower Creeks between 1716 and 1718 (Boyd, 1949, 1952; Fairbanks, 1974, 1978; Sturtevant, 1971) and opened a small trading post in 1738 along the St. Marks River, establishing a second trading post in 1745 to further entice the proto-Seminole to engage with the Spanish (TePaske, 1964: 215–218). They had earlier reestablished a military presence near the St. Marks River, which clearly was presented as an enticement for the Creeks to come to Florida (Boyd, 1949, 1952)—remnant Apalachee specifically (Barcia, 1951: 366). Living within the realm of Spain had other benefits. There were fewer Spaniards, they maintained less control of their colony, they interfered less with indigenous trading practices, and they were easier to manipulate to supply goods (Sattler, 1996: 50). Nonetheless, Swanton (1922: 101) characterized the Spanish as only "passive sympathizers." They did not work hard to revitalize their province, and the "pull" mechanism offered by Spain should not be considered to overshadow other factors intrinsic to indigenous society itself (see below).

The environment of Florida is also viewed as an attractor, Apalachee province in particular (see Reding, 1935: 169, 170, and also the discussion of Apalachee agriculture in Hann, 1988). There were large herds of wild cattle and horses, prime hunting environments (Boyd, 1949: 15, 18), and a soils geology similar to that of central Georgia. The earliest proto-Seminole presence was located near water bodies with associated hammocks that encouraged agricultural activities (Sattler, 1987: 23; Fairbanks, 1978) and a generalized mixed subsistence economy during a time when resiliency and flexibility were paramount to survival (Craig and Peebles, 1974). Hunting trips back into the peninsula may have preceded actual permanent migration and resettlement (Wright, 1986: 63–64), thus providing an explanation for how the proto-Seminole became aware of the opportunities afforded by the old Spanish mission fields.

These factors are treated as received wisdom, yet they treat the earliest proto-Seminole communities as monolithic and assume that these stimuli were experienced uniformly or embraced widely. However, the separate proto-Seminole regions were settled by distinct groups of people speaking different languages and at different times (see figure 7.1). And it took many decades and generations for these people to coalesce; and even then, and still to this day, it would be incorrect to say this common Seminole identity is embraced by all (see Wright, 1986: 319). The two main geographical divisions were the Apalachee old fields in the area of modern-day Tallahassee (Sattler's Apalachee Seminole) and the Alachua prairie in the former Timucua district near modern-day Gainesville (Sattler's Alachua Seminole).

Outlining the timing of these founding events is critical to the interpretation of Seminole ethnogenesis presented in this chapter. Although there is some disagreement in the literature about when the Seminole came to Florida, this largely reflects the bias toward the Alachua Seminole in earlier research, about whom much more is known (see Cline, 1974; Craig and Peebles, 1974). In other words, early scholars of the "Seminole" were speaking of the Alachua Seminole because historically the name was first applied to these bands (see Swanton, 1922). Another factor adding to the confusion is the earlier conflation of the two founders of the principal Seminole settlements: Secoffee in the west and Cowkeeper in the east (Cline, 1974: 131–132; Porter, 1949; Swanton, 1922: 398–399). They were not the same person.

Others have proposed that the Apalachee Seminole who settled around

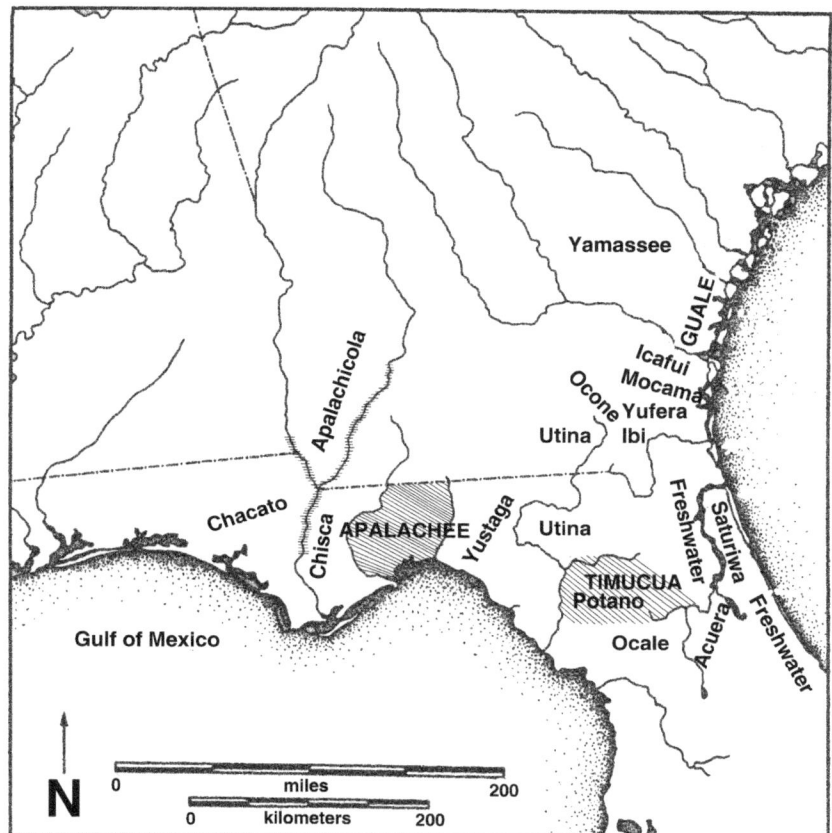

Figure 7.1. Map of early proto-Seminole settlement locations circa 1720–1784. Shaded areas represent those areas associated with the Apalachee Seminole and Alachua Seminole. Early Seminole settlements were also located along the Flint and Chattahoochee rivers. Figure modified after Sattler 1987: 22.

Tallahassee were the first bands to make the move permanently. Sturtevant (1971: 101) places this migration around 1727, while the Alachua Seminole migrated sometime during the 1740s or 1750s (Sattler, 1996: 48). Although the Spanish sent an envoy to the Lower Creeks between 1716 and 1718 (Diego Peña) to entice them to resettle Apalachee province (Boyd, 1949, 1952), and they agreed to the move (Boyd, 1952: 123), it is unknown when people acted upon these promises. As further confirmation of the timing of the migrations, Sattler (1996: 46) reports two different Seminole traditions that indicate that the Apalachee Seminole were the first to migrate to Florida; in his analysis these are more credible because they are reported by 19th-century

Seminole groups not affiliated with the Tallahassee bands. Sattler also interprets the statement of Tonapi (Tunapé), the leader of the Apalachee Seminole in the later 18th century, as suggesting he moved to Florida between 1724 and 1740.[9] Based on these data, Sattler (1996: 47) concludes that "the Apalachi and Yamasi first settled the Apalachi region beginning in 1718. The main emigration, however, apparently occurred in the 1720s and 1730s, involving Simpukasi [Secoffee] and his followers." Although this interpretation of the timeline is critical to the perspective developed in this chapter, it is not universally reported in Seminole histories.

The specific association between distinct ethnic groups and the early proto-Seminole settlement areas is also important (see figure 7.2). The Alachua prairie was settled predominantly by ethnic Oconees under the leadership of Cowkeeper. They were pro-British and likely left the suzerainty of Brims because of his vacillating approach to colonial diplomacy. Although other ethnic groups were incorporated into the Alachua Seminole, including ethnic Yamassee (Sattler, 1996: 48), the core was composed of Oconees and other Lower Creeks. Sturtevant (1971) notes that the Yamassee were actually slaves, although Cowkeeper, their leader, was married to a Yamassee woman (Swanton, 1922: 399). Nonetheless, despite their anti-Spanish tendencies, the Oconee Creeks "spoke Spanish, and seemed to follow Spanish customs" (Cline, 1974: 123), an intriguing statement not pursued further. Bartram (1988: 164) noted cultural elements similarly "tinctured with Spanish civilization" with "Christians among them." It should be clear, however, that the Alachua and Apalachee Seminole were quite antagonistic toward each during much of the 18th century.

The Apalachee Seminole were pro-Spanish, Hitchiti-speaking bands founded under the leadership of Secoffee, the kin of Brims of Coweta.[10] Secoffee was succeeded by Tonapi, also of Coweta and probably related to Secoffee in some way (Sattler, 1996: 46). Both were very pro-Spanish (see Fitch, 1916: 184); Secoffee may have been baptized (Hahn, 2004: 136), and in 1717 Secoffee and his kin Chilokilichi visited St. Augustine to swear allegiance to the Spanish Crown (Barcia, 1951; Boyd, 1952). Chilokilichi established a town at the forks of the Flint and Apalachicola Rivers in 1715 where the pro-Spanish faction of the Lower Creeks could find respite from the neutralist policies of Brims (Crane, 1956: 134). This town had a large contingent of ethnic Apalachee living in it (Barcia, 1951: 366). Secoffee and Chilokilichi were joined by chief Adrian of Bacuqua, an Apalachee chief who survived the Moore raids (Boyd, 1952). Secoffee himself was half-Apalachee and he was

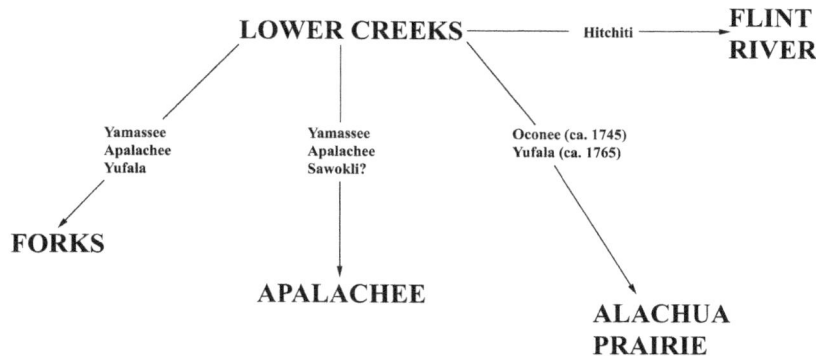

Figure 7.2. Ethnic realignments during the early 18th century (after Sattler 1987).

also married to an Apalachee woman, as apparently was his kinsman, Brims of Coweta (Barcia, 1951: 362; Hann, 1996b: 67). As noted by Fairbanks (1978: 164–165), "whether these wives were Florida refugees from Moore's raid of 1704 is not known, although many Apalachee fled to the Creeks for protection." Fugitivism was certainly common throughout the Spanish mission period and it is possible that these Apalachee had left Spanish territory before the missions were destroyed. Interestingly, Fairbanks continued, "neither is it possible to evaluate how much this involvement of influential Creeks with Apalachee may have been responsible for the subsequent movement of other Creeks into Florida. It is certain that Brims's heirs maintained a strong interest in the area" (Fairbanks, 1978: 165).

Sattler (1987, 1996) pushes this connection between the Apalachee and proto-Seminoles even further. His analysis and accounting of early 18th-century travel narratives and official reports indicates that the Apalachee maintained a sizable presence throughout the region during the early 18th century (Barcia, 1951: 358, 359; Boyd, 1949: 22; 1952: 109, 116; Boyd and Latorre, 1953: 114). They maintained separate villages among the main Lower Creek towns, were incorporated into some Lower Creek towns, and maintained a presence in areas distinct from the Lower Creeks throughout peninsular Florida. Other ethnic identities were also present in Lower Creek towns in the early 18th century, including Christian Yamassee, Timucua, and Mocama, as Diego Peña and William Bartram reported (Bartram, 1988: 307; Boyd, 1949: 25–26; Cline, 1974: 64; Swanton, 1922: 106).

Although the Timucua and Mocama are rarely identified after 1706 (but see Bartram, 1988: 307; Boyd, 1949: 26), the Yamassee, who suddenly appeared in Spanish Florida in the latter half of the 17th century, are much better known historically (Green et al., 2002). They were a highly ephemeral yet ubiquitous group that was found in pockets throughout the Southeast during the 18th century (Swanton, 1922; Worth, 1995). They were originally thought to derive from Georgia, perhaps subject to the Guale cacique during the Spanish era or connected to the Ocute (Hitchiti) during the 16th century (see Crane, 1956: 164, for a Cherokee origin myth). Swanton (1922: 82) considers them closely allied with the Guale, and the two ethnic groups may have merged throughout the 17th century. The Yamassee experienced a complex relationship with the Lower Creek leadership during and after the Yamassee War of 1715 (Oatis, 2008). Previously complicit with Brims during the war, the pro-British Coweta Creeks were quick to turn on the Yamassee; a fact that caused considerable internal turmoil among Lower Creek leadership. Retaliation against the Yamassee was complicated because "for as Shure as we Kill A Yamassee, he has a Relation or freind [sic] amonge The Creek's" (Fitch, 1916: 182). As Sattler (1987, 1996) has noted, the Yamassee were well integrated into Lower Creek communities; they had a village in Apalachee province during the 1730s, requested friars to serve them, and may be ancestral to the Oklawaha band of Seminole (Swanton, 1922: 107).

The annals indicate the Apalachee were distinctly interested in returning to their homeland and pushed for the construction of Fort San Marcos to offer protection for them to do so (Barcia, 1951: 366; Boyd, 1952: 112). The principals involved in this effort were the aforementioned Secoffee, Chilokilichi, and Adrian of Bacuqua (Boyd, 1952: 112–113). The request was made so that "the Apalache Indians might descend from the hills onto the plains and settle down in the assurance that their enemies could not hurt them." In short, they wanted to return home and to the protection of the Spanish where they could continue a Catholic life, receive Catholic instruction, and baptize their children (Boyd and Latorre, 1953: 100, 107–108, 114–115). The desire of former Christian Indians to return to the lands of the Spanish is reflected in the number of villages willing to commit, as Diego Peña reported:

> he [Secoffee] (will) move to the Chicazas[11] of Apalachee, that (he) is expecting a hundred followers who are in the other province. . . . The villages which are to move are Tasquique nation [Yamassee], and in addition Apalachicolo, Sabacola, (that) of Chislacasliche [Chilokilichi], and the chief of Bacuqua, with two men, and it is said with perhaps more, and the said Uchises. The Apalacheans continue

dispersed, the reason for which, say the English, (is) that they and the Yamassee do not wish to exercise vengeance (Boyd, 1952: 123).

If, as has been suggested (see Boyd, 1949: 26; 1952: 134), Tasquique is affiliated with the Yamassee, then this village likely had a Spanish or Christian element residing there. The Apalachicolo and Sabacola also had gained some experience with Catholicism during the late 17th century, and the latter's village was noted as containing Christians (Boyd, 1952: 134; Hann, 1996b: 67). Finally, Bacuqua was an actual Christian Apalachee village during the 17th century.

The common thread among these villages is their pro-Spanish leanings and possible connections with Christian Indians. When these villages actually moved is subject to some debate. But the move did occur, and it did so at the wish of a pro-Spanish Apalachee affiliated leadership in the context of disparate groups of Apalachee, Yamassee, and Timucua Christians, among others, who had survived Moore's raid. Because these migrations occurred prior to those that established the Alachua Seminole presence in central Florida, we can say that these efforts were the beginning of the Seminole people.

However, this was not a random series of migrations that were divorced from larger cultural patterns (Sattler, 1996: 62; Swanton, 1922: 400). This was an active decision on the part of the people involved, and in this regard the leadership of the communities was paramount. It was the elites who maintained control of the sacred fire and its *hiliswa* (medicine). These individuals had the legitimacy to establish new towns because of their inherited status (Sattler, 1987, 1996), and the Spanish were eager to court their presence in Florida, seeking their consent to move with the promise of land (Hahn, 2004: 146). However, the rank and file was also not a random group of people that slowly leaked into Florida because of hunting prospects or disaffection with the English. The proto-Seminole sought to bring with them relatives and friends from a diverse number of villages and to recapture those who had been taken as slaves during the Moore raids (Barcia, 1951: 361, 369, 370). The entire process was very corporate, highly structured, and layered with intention. Seminole ethnogenesis was a deliberate process from the start.

And this is why the biological data presented in chapter 1 are so important. The interpretation of changing patterns of gene flow during the 17th century suggests that the Spanish-allied communities were actively forging a common social identity with considerable flexibility and diffuse rules of membership. Increased gene flow among disparate identity groups resulted in the conflation of social categories as Apalachee, Guale, Timucua, and Yamassee populations experienced an increasingly similar life course experience throughout the 17th century. It was this burgeoning sense of a "Spanish Catholic Indian" identity that was responsible for the earliest proto-Seminole migrations back to Florida—and which I propose is of ultimate importance for understanding the initial stages of Seminole ethnogenesis. Without these social linkages—the ties of identity that innervated the Catholic communities and beyond (see chapter 8)—social adaptation to the changing political environment of the Southeast would have been much more difficult. Shared identities represented bonds of kinship and affinity that stretched over dozens of villages, and it was the flexibility of residency, while maintaining the status of specific identities, that initiated Seminole emergence. The details presented in chapters 4–6 indicate that conditions were rife with opportunities for ethnogenesis to occur among these disparate communities. The data presented in chapter 1, interpreted within the social theoretical framework presented in chapters 2 and 3, indicate that people responded to these conditions in a manner that led to ethnic homogenization.

In this regard, the Apalachee hold an important position. They were feared, well known, and highly respected by their contemporaries from the 16th century onward. They were the most numerous constituent of the Spanish provinces during the early 18th century, and as such, they were widely dispersed by the joint Moore/Creek assaults on the Florida missions. Apalachee women had intermarried with key Lower Creek political leaders (as had Christian Yamassee; see Hahn, 2004: 132; Swanton, 1922: 399). And these facts explain why the Apalachee region was the first to be settled by the proto-Seminole—the Apalachee *were* the proto-Seminole. But they were not alone. This was a polyethnic group, initially speaking multiple languages but quickly finding a common tongue in Hitchiti. The ethnogenesis of the Seminoles began as a very conscious decision to return to their homeland and bring with them those ethnic "others" with whom they had forged a broader basis of affinity in preceding generations. But it was the Apalachee attachment to the landscape and their sense of ethnic pride (as this book's opening epigraph indicates) that promoted the move to Florida to begin with.

Ethnogenetic processes defined a pan-La Florida identity that went beyond just Apalachee attachments, however. The developing sense of flexibility—that is, liminality—among specific communities provided the social networks needed to survive difficult times and to act opportunistically and with agency as circumstance prescribed. And it was this shared sense of identity that helped promote and operationalize the move to Florida that begins the process of Seminole-Creek divergence. The proto-Seminole did not come back to Florida because they were pushed by the English or pulled by the Spanish or a promise of better hunting or agricultural lands. They migrated to Florida principally because they actively chose to do so. Such mobility and flexibility was part of the normal course of colonial life for the southeastern tribes, responding to opportunities and changing allegiances as best suited them.

However, although the Apalachee and Alachua groups would both ultimately become Seminoles, it is important to emphasize that the original migration rationale was quite distinct for each region. Although the Oconee may have had some earlier experience with Christianity,[12] their presence in central Florida during the mid-18th century may directly relate to the existing presence of the Apalachee Seminole in the panhandle. As Sattler notes,

> throughout the 1730s and 1740s, the Okoni and other Creek groups raided in this area [northern and central Florida], harassing the Spanish and taking the Yamasi and other Spanish-allied Indians prisoner. . . . During the 1740s and by 1750, these frequent raids into Florida led to the establishment of permanent settlements by the Okoni in the area of the Alachua Prairies, near modern Gainesville. (Sattler, 1987: 19–20)

Although the allure of central Floridian hunting and pastoral opportunities may have brought them to Florida anyway, the Oconees' knowledge of the subsistence possibilities can be attributed to the continued practice of frontier raiding that dominated Creek-Florida interactions during the last decades of the 17th century. If the Apalachee Seminole and other Christian Indians had not returned to Florida as the earliest wave of proto-Seminole, the Alachua bands might have had no reason to be in Florida to begin with.

Regardless of the interaction dynamic among the Apalachees, Oconees, Timucuas, Yamassees, and Guales during the 17th century, there may have been another reason that the Oconee, like many other ethnic groups, ended up as Seminoles in Florida rather than as Creeks in Georgia. This relates to their shared status as ethnic minorities within the Creek political ethos, a distinction that intensified around the turn of the 18th century due to the widespread effects of epidemic disease in indigenous communities removed from direct European contact.

Motivation and Rationale: The Effect of Indigenous Ethnic Rivalry

Throughout the historical annals the relationships among the English, Spanish, and various Creek factions are central to the narrative but read as somewhat confused, as if the entire subtext of the negotiations was not known by the European recorder (e.g., Fitch, 1916; Taitt, 1916). Indeed, this is likely the case. However, historian John Leitch Wright (1986) has provided a novel perspective on Creek and Seminole history which provides clarity to the muddled internal Creek politics and which further ties the interpretation of Seminole history presented in this chapter to the data presented in chapter 1.

Wright picks up on a well-reported and well-supported supposition about early Creek history. Oral tradition suggests that the principal founding towns of the Creeks represent the intrusive presence of Muskogees who, legend says, migrated to Georgia and Alabama from somewhere west of the Mississippi (Gatschet, 1969; Hahn, 2006; Hann, 1996b: 70; Sattler, 1996: 42; Wright, 1986: 8, 156). The pure Muskogee were residents of the principal northern Muskogee Creek towns of Coweta, Cussita, and Abeika (among others), and certainly smaller ethnic groups were incorporated into the dominant Muskogee ethnicity through intermarriage, conquest, or aggregation. They spoke Muskogee and their individual ethnic identities were maintained to some extent despite objectification (by the English) as Creeks. These individuals and the towns they lived in formed the core of the Uchises (a term applied to them by the indigenous Georgian and Alabaman ethnic groups), later to be called Creeks by the English who glossed over the degree of ethnic complexity represented. These individuals spoke a language distinct from other populations in the region, and they were "pure Muskogee" in their eyes.

However, as Wright (1986: 3) noted, "most Creeks did not speak Muskogee, at least not as their first language, and frequently not at all." For that matter they did not consider themselves Creeks, in the English view of the world. Those populations resident in Georgia, Alabama, and Florida at the time of this migration, as well as all of the other peoples incorporated into the Creek political body during the colonial period, were not pure Muskogee. These people were a diverse collection of ethnic groups defined by what they were not. Rather, they were

> Hitchitis and refugees such as the Koasatis, Alabamas, Tuckabatchees, Yuchis, and Shawnees who were not absorbed by the Muskogees. Among the Hitchitis—

"the original inhabitants"—and those speaking closely related dialects were Chiahas, Apalachicolas, Tamathlis, Sawoklis, Yamacraws, Yamasees, Guales, Oconees, Osochis, Miccosukees, and possibly Timucuas. (Wright, 1986: 16)

To this, of course, can be added the Apalachees (Wright, 1986: 13). These peoples were collectively "stinkards," a lower class of people, inferiors in the eyes of the politically dominant Muskogees. Although members of these subordinate ethnic groups were often absorbed into the Muskogee ethos, many were not and remained minority refugee elements within the burgeoning Creek politic. Wright proposes that European contact heightened these divisions as this dual ethnic moiety system assumed a more important ordering dynamic in their lives.

Although all scholars of Creek and Seminole history note the ethnic diversity within Creek villages, Wright emphasizes these internal ethnic differences and (re)interprets almost all of Creek and Seminole history within the framework of a dual ethnicity or dual moiety system of Creek sociopolitical structure. In Wright's analysis, this structuring organization was so pervasive and important that it replaced the white/red, peace/war distinction at the town (not clan) level and helps explain major events in Creek history from the perspective of the participants rather than the observers. The implications are profound. These ethnic tensions between pure Muskogee and non-Muskogee affected major events in Muscogulge history and "helps explain events in the East before western removal, why more and more Muscogulges became known as Seminoles, and why after 1861 Tuckabatchees sought refuge in Kansas and Cowetas (Muskogees) in Texas when they were not shooting at each other in Indian territory" (Wright, 1986: 18). Interestingly, the issue of ultimate causation of Seminole origins is not directly addressed by Wright (1986). In fact, there is only a single entry for "Apalachee" in the index.

Nonetheless, the critical point is that the proto-Seminole were bands of these minority, remnant ethnic elements; the stinkards afforded an inferior position within the sociopolitical world of the 18th-century Southeast. It was the Apalachicolas, Apalachees, Guales, Timucuas, Yamassees, and Oconees that first made the move to Florida. And Wright's analysis provides an emic reason for this. They were outcast by the dominant Muskogees and treated as inferiors. Some were pro-Spanish, others were pro-English, but they were all not pure Muskogee. For the leadership of the proto-Seminole, those with the right to establish new towns, moving to Florida ameliorated their subordinate position within the Creek political body (Sattler, 1996: 63). As such, the dual ethnic moiety model provides a powerful emic rationale for proto-Seminole emigrations and does so without any explicit appeal to exogenous European influences. Seminole ethnogenesis is, therefore, provided another facet of intent. The Christian elements wanted to return to Florida because Apalachee province was their homeland, and the non-Muskogee leaders could seek political advancement independent of Brims's authority. The Oconees (who founded the Alachua Seminole), however, may have been true foreigners in Florida (but see above) but shared with the Apalachee and Apalachicola Seminole their inferior status within Uchise political discourse.

The dual ethnic moiety model articulates well with the perspective developed in this book. Although the bioarchaeological database was limited to the Spanish provinces proper, these non-Muskogees buried in the mission churches (Apalachee, Guale, Timucua, and perhaps Yamassee) experienced ethnogenesis due in large part to the same dynamic between Muskogee and non-Muskogee. It was the objectification of the Apalachee, Timucua, and Guale during the latter half of the 17th century that initiated the liminal stage of ethnogenetic transformation among these communities (chapter 5)—objectification at the hands of the English through their allies in central Georgia: the proto-Creeks. In this sense, the ethnic dialogue that defined a dual moiety system in Wright's analysis extends back into the 17th century and explains much more than just Creek and Seminole 18th- and 19th-century history. It explains the ultimate basis for how these distinct identities evolved to begin with. And we can observe in the patterns of biological variation the incipient emergence of this new social identity among the Florida and coastal Georgia communities.

It is important to note that this interpretation is based only on data from Spanish missions. That is, this focus on mission church contexts biases the database toward Spain's faithful and ignores the fugitives, transients, and refugees who made an appearance in Florida during the latter half of the 17th century. However, these refugee populations were also ethnic minorities within the universe of Muskogee sociopolitical discourse. During the late mission period (1650–1700), these peoples descended on Apalachee province, which experienced population aggregation as disparate ethnic communities were trying to escape the slave raiders operating in central Georgia. These people were Chines, Chacatos, Pacaras, Amacanos (Yamassees), Tamas, Tocobagas, and perhaps Oconees. And population size data suggest 8–10 percent of the population of Apalachee province was composed of these minority elements during the last quarter of the 17th century (Hann, 1988: 102, 103, 165, 179, 322; 2003: 121–122). Sometimes these people converted to Christianity but sometimes they did not. They generally maintained separate villages within the jurisdiction of the Apalachee chiefs but it is likely that gene flow occurred among these communities.[13] In other words, they may have also participated in the process of ethnogenetic transformation because all of these ethnic groups would have suffered objectification by the Muskogee Creeks.

When the missions were destroyed these communities were subject to the same diaspora as the resident Catholic populations. Those who fled to the Chattahoochee River (the other significant region of population aggregation) remained minority, inferior elements within Muskogee society. And when the time was right, they used their burgeoning sense of shared identity to take action. They came back to Florida, joined by other disaffected peoples who all eventually became known to the Spanish as *cimarrones*, to the English as Siminioles, and to us as the Seminole.

The turning point for the Creeks and Seminoles may have been the smallpox epidemic of 1696, which Kelton (2007) identifies as the first pan-Southeast pandemic to affect a majority of interior and coastal populations. While it is undeniable that epidemics affected the earliest Spanish and English colonies along the east coast, there is a growing tendency to view these as more localized phenomena rather than as widespread pandemics spreading over hundreds of miles (Henige, 1986; Kelton, 2002, 2007; Ramenofsky et al., 2003; Stojanowski, 2005a). However, the 1696 epidemic, and those that followed in quick succession, shattered many communities and sent myriad refugees into the Creek heartland. The slow leaking of populations from Spanish Florida into central Georgia during the 17th century was nothing compared to the rate of aggregation during the 18th century. And it was just this sort of diaspora and coalescence that would have initiated the internal ethnic dialogue that Wright (1986) proposes. As more and more people of diverse ethnic backgrounds congregated in the Georgia and Alabama interior, internal ethnic tensions may have fomented, leading to, among other things, a return to Florida by those populations suffering objectification as ethnic others.

When interpreted this way, there is clearly a historical continuum linking the 17th and 18th century, a continuum identified through the dynamic of inter-Indian ethnic rivalry. There is also a biological continuum through the recognition that many of the nonpure Muskogees living in Georgia during the 18th century were disaffected Florida and coastal Georgia fugitives who were escaping the hardships and impositions of the Crown and the friars. And then there is a biosocial continuum when these two perspectives are combined and interpreted within the purview of the migration rationale of the earliest proto-Seminole communities. It was the distinction between Muskogees and non-Muskogees that defined the 17th-century interaction between La Florida communities and those living in central Georgia (chapter 5). It was the exact same distinction that defined similar geopolitical arrangements during the 18th century, whether one speaks of Creeks and Seminoles or Apalachees, Guales, Timucuas, Yamassees, or Oconees.

What does it mean that ethnogenesis was occurring among Florida's indigenous communities if the process was ultimately preempted by the English, and La Florida's populations were killed, enslaved, or scattered throughout the Southeast? This chapter argues that changes in population structure throughout the 17th century are not just of academic interest. Rather, these data provide a unique biological and evolutionary backdrop against which historical and archaeological data can be interpreted. Some of the earliest "Seminole" to return to Florida settled in Apalachee province, the leaders of these bands were married to Apalachee women or had Apalachee affiliations, and the Apalachee were the majority ethnic group in La Florida during the later 17th century. As such, they were widely dispersed when the missions were destroyed. The Apalachee had a strong sense of identity and connection to the landscape, a connection that positions incipient Seminole ethnogenesis as a return to their homeland. But these individuals were joined by a much broader constituency defined through the active negotiation of ethnic boundaries prior to their diaspora and subsequent nomocide leading to historical invisibility (figure 7.3).

The model of Seminole ethnogenesis presented here suggests that the ultimate cause of Lower Creek-Seminole divergence was increasing social and political distance that developed when distinct communities of "Creeks" left central Georgia to settle in the old mission fields of La Florida. The rationale for this emigration was related to internal Creek ethnic discourses as well as negotiations and shifting allegiances among Spain and England, and between each of these nations and the indigenous allies they sought to manipulate. But the biological background to these migrations, as only observable through biodistance analyses, provides a unique subtext to the historical narrative (figure 7.4). This biological perspective pushes back the

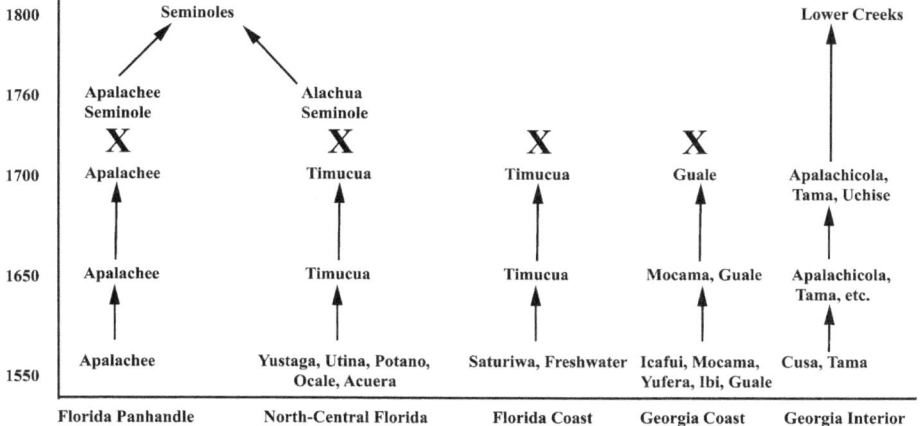

Figure 7.3. Pattern of etic (ascribed) social identities associated with specific regions of Florida and Georgia. Vertical axis represents time. Large X icons indicate the cessation of that identity in the historical annals, which represent cases of nomocide.

beginnings of Seminole history into the 17th century when these intercommunity connections, which implemented proto-Seminole emigrations, were being forged. This biosocial model emphasizes the importance of a continuing historical process related to negotiations of ethnic identity among indigenous communities, the practice of ethnic soldiering, and the flexibility of Mississippian cosmogony that allowed for shifting allegiance, social identity, and residence "without being destroyed in the process" (Wickman, 1999: 16). Therefore, this perspective effectively links the Spanish and post-Spanish (Seminole) periods of Florida indigenous history.

However, things are a bit more complicated than this. The argument laid out in this chapter, while it does find support in the work of several historians, rests on a very specific sequence of events and inferred motivations for the proto-Seminole migrations whose exact dates cannot even be determined. While the distinction between the Alachua and Apalachee Seminole has been highlighted, and the earlier migration of the latter to Florida has been emphasized, there is really no reason to assume that the Florida peninsula would have remained vacant throughout the 18th century if Christian, proto-Seminole populations had not returned to Florida. Or is there?

On one hand, one could argue that historical speculation is meaningless. The facts are that the Apalachee Seminole did return to Florida. And despite the complexity of the later history of the Seminole, this nucleus may

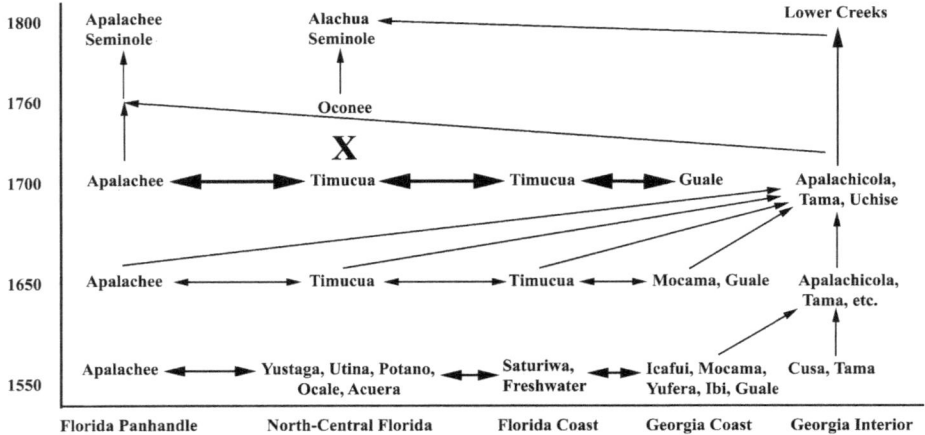

Figure 7.4. Patterns of biological integration associated with changes in social identity categories as applied to indigenous groups in each region. Horizontal double arrows represent patterns of gene flow among mission populations as inferred from analyses presented in chapter 1. The length of the arrow indicates the strength of integration between populations. Diagonal arrows represent interregional patterns of migration and/or fugitivism.

have seeded the peninsula and acted as a magnet for subsequent migrations of Lower Creek discontents over the next 100 years. That some indigenous presence once again attracted peoples back into Florida from Georgia is not entirely unreasonable. Based on 17th-century reports, it is possible that these very same people had relatives who had, a generation before, left Florida for Georgia, and before that had gone to Georgia from Florida. The record of fugitivism and transiency during the 17th century negates the biological reality of the modern state border. This notion of "seeding" applies a positive pull mechanism for subsequent movement of peoples back into Florida once the proto-Seminole territories were established. In addition, historical records indicate that the proto-Seminole continued to serve as "target populations" caught between England and Spain. They were subjected to raiding during the 18th century as they had been during the 17th century. This fact applies a negative pull mechanism to later Seminole emigration that does, however, establish a degree of behavioral continuity that links the 17th and 18th centuries. In either case, once people came back to Florida, for whatever reason, Seminole ethnogenesis commenced as history suggests.

On the other hand, the continued interaction between Lower Creeks and the English and Spanish governments almost guarantees that some bands would have settled in Florida eventually under any circumstances and regardless of whether the Christian, pro-Spanish communities forged the return to Apalachee province. They would have acted as a nucleus of aggregation for other Lower Creek dissidents and would have provided a reasoning for Oconees (acting as ethnic soldiers) to explore north-central Florida while raiding these proto-Seminole communities. While quite possibly true, this perspective ignores the facts of history and is somewhat nihilistic as such. It relegates Seminole ethnogenesis to a forgone conclusion that, in addition to being antihistorical, completely deflates the primordial charter (that is, as untamed resistors) of the Seminole peoples.

Regardless of the issue of temporal primacy or historical speculation, emphasis on the importance of Spanish-allied Christians throughout this chapter belies the inherent bias that appears throughout biohistorical research: the naming and reification of populations as part of the reconstruction of a linear history. If the Apalachee and other Christian populations had not returned to Florida to become Seminoles, would it be appropriate to then declare the Seminoles "adopted children of Florida" (Wickman, 1999: 18)? To affirm this assumes that Florida and Georgia tribes existed in a state of monolithic seclusion in prehistory, a perspective eschewed by proponents of ethnogenetic theory (see chapter 3). Indeed mission period fugitivism was a problem throughout the 17th century. As such, it is likely that many "Creeks" in central Georgia were actually disaffected Spanish mission Indians. And if dissident Creeks then migrated to Florida to become Seminoles, the entire framework of discussion falls apart from the perspective of tracking biological lineages for the purpose of establishing "legitimacy." Unfortunately, I currently lack contemporary 17th-century data from the Georgia interior that would allow consideration of the degree of biological integration among Florida and Georgia populations during the mission period. However, skeletal data do exist to evaluate patterns of integration during prehistory, and in the final chapter the timeline of Seminole ethnogenesis is pushed back even further into the protohistoric period.

Eight *Back to the Past*

Precontact Biological Integration as
Prelude to Colonial Ethnogenesis

> In the face of the overwhelming evidence to the contrary, no objective observer now believes that the so-called Seminole or their ancestors, the Creeks, were aboriginal dwellers of Florida or had been there, as they sometimes claim, "since time immemorial." The historic "Seminole" are descendants of emigrants, those Indians who repopulated Florida in the eighteenth century. The time, place, and circumstance of their establishment in Florida are matters of historical record. The sequences are incomplete only in minor detail, but unquestionable in main outline.
> —Howard Francis Cline, *Notes on Colonial Indians and Communities in Florida, 1700–1821* (1974: 91)

Patricia Wickman's analysis of 16th- through 18th-century indigenous history focuses on the cosmogony of, in her words, the Maskókî peoples: "tens and hundreds—perhaps thousands—of individual tribes, networked together in cultural kinship by greater and lesser elements of a coherent cosmogony" (Wickman, 1999: 210). This cosmogony eschewed a linear view of history and assumed a circular form based on flexibility and dynamism, freedom of movement and residence, a social organization that allowed connections to be made between peoples such that precontact trade and exchange, patterns of warfare and elite exogamy (see chapter 4) effected a multilingual and polyglot series of villages rather than discrete, isolated tribes. Such a perspective forms the core of the ethnogenetic critique of anthropological historical analysis, as noted in chapter 3 of this volume. Through her analysis, Wickman concurs with the view of Sattler (1987, 1996) about the identity of the earliest Seminole communities. They were composed of Apalachees, Timucuas, Guales, and Yamassees who had previously been living along the Florida mission chain during the 16th and 17th centuries. Indeed, such connections to the ancients in Florida are affirmed by modern Seminole oral histories:

> The cultural repertoire of today's Seminoles and Miccosukees holds strong memories of Hitchiti, Yuchî, and Yamásî ancestors and songs from the Calusa. At least one family honors memories of tall Abalachi women as Clan Mothers. Numerous Oklahoma Seminoles preserve a Spanish heritage as a part of their tradition. And Maskókî peoples in both Florida and Oklahoma support an active, ongoing kinship system based partially on the relationships of core Clans and "found people." (Wickman, 1999: 18)

However, Wickman goes a step further in questioning the perceived distinction between the various tribal groups recorded in the pages of history. Although the tribal ethnonyms in many cases reflected part of the sociocultural "reality" of southeastern U.S. political discourse, the imposition of a tribal model of isolation and discontinuity on what amounted to a complex interaction system creates a false sense of history. And this perception of indigenous community structure in large part reflects inherent biases imposed by colonial period European observers read and interpreted by modern Europeans with a similar worldview.

Wickman positions within the matrix of Maskókî identity a series of propositions about the Spanish and English colonialist agendas and biases, and these propositions have gained greater traction in recent literature. The first of these is the myth of total power, that the Europeans dominated indigenous tribes due in large part to their technological superiority. The issue of firearms was discussed briefly in chapter 5; however, this myth does not focus on technology alone but on the mythology of the "Conquest" in modern historical narratives. The second myth is that of total victory, which also casts the colonial period in terms of a "Conquest" approach to history. However, the belief that Native Americans were simply overwhelmed and suffered utter defeat is antithetical to the very existence of the fields of colonial history and historical archaeology that seek to reconstruct the decades and centuries of negotiations that occurred between indigenous and European entities. The third myth is that of total destruction—that Native American populations were systematically and uniformly destroyed by European technologies and the diseases Europeans unwittingly brought with them. This last point is the most germane here.

Earlier viewpoints on epidemic mortality that proposed rapid annihilation over long distances (e.g., Dobyns, 1983, 1991) are being replaced with a more measured assessment that focuses on the patchwork of epidemics that affected specific communities (see Henige, 1998; Kelton, 2002, 2007; Ramenofsky, 1991; Stojanowski, 2005a). To counter the historical assessments of population sizes (which demonstrate a measurable decline through time), scholars are questioning both the basis of the initial population size estimates and the biases inherent in the records of decline that followed. Precontact population size estimates are largely unknowable and based on reconstructed estimates from often vague and difficult to interpret passages in the chronicles, something Henige (1998) has most vocally criticized. When the baseline is suspect, then all subsequent assessments of demographic collapse are equally suspect. This is not to deny the presence of epidemics which are recorded in the 16th century and certainly devastated indigenous communities. However, other factors were contributing to demographic collapse, and these tend to be diminished in lay discussions of the colonial period (see Worth, 1998b, for a discussion of these other mechanisms of demographic collapse).

One of these alternative factors was fugitivism. People simply left the Spanish system for better lands and more favorable living conditions. And as noted by Wickman, it was the Maskókî cosmogony which allowed Apalachees, Timucuas, and Guales to "quit" the mission system, as the Yamassees so clearly and visibly did during the 17th century. Their worldview, which was shared widely throughout the Southeast, allowed them to join other villages and communities, and in response to depredations many Christians simply left the Spanish sphere of influence to destinations north and south of the *camino real*. In their absence they were not counted in colonial enumerations and disappeared into the villages they migrated to, where new ethnic identities were applied to them. There is, of course, ample historical evidence for mission out-migration (Wickman, 1999: 185–186; Worth, 1995, 1998b) as the Spanish accounts discuss the recovery of these fugitive runaways, these *cimarrones*. In this sense, the precontact Floridians were not only directly ancestral to the Seminoles (chapter 7) but were also incorporated into villages in Georgia that would become known as the Creeks.

However, a more subtle point is to be made here. Even where recognized as a significant component of demographic collapse, fugitivism is unintentionally presented as a reactionary behavior to epidemic disease and the rigors of mission life and Spanish imposition. However, in Wickman's analysis, the *cimarrones* were not just "quitting" Christianity because of the Spanish but because this is what they had been doing all along. The tribal and chiefly identities that appear in the protohistoric chronicles were the products of a regional Mississippian interaction sphere inherently subject to population redistribution (see chapter 4). And when Spanish Catholics left their natal villages due to disaffection they may have followed preexisting patterns of migration, patterns that should be visible during the precontact period and that defy the very logic of casting the Seminole as derivative migrants to Florida regardless of which proto-Seminole communities were settled first, by whom, and the rationale behind these relocations (see chapter 7).

To close this volume, I therefore return to the realm of biodistance analysis to consider a broader swath of the Southeast, one that includes the interior populations of Georgia. In chapter 1, I focused exclusively on populations derived from areas subjected to Spanish missionization. The reasoning was well founded. By normalizing the geographic area sampled through three successive time periods (precontact, early mission, late mission), I could make estimates of regional genetic variability (F_{ST}) that were not biased by geographical scale. However, by including data from the interior of Georgia it is possible to evaluate how well integrated the Florida and interior Georgia communities were *prior to European contact*. In this chapter I analyze the pattern of phenotypic variation present in Florida and Georgia during three periods of prehistory: the early precontact period, the late precontact period, and the protohistoric period.

Research Design

The distribution of precontact skeletal samples is presented in figure 8.1. Thirty-three samples are included in this analysis that date to three time periods. The early precontact period (A.D. 500–1100) includes samples from nonagricultural populations, the late precontact period (A.D. 1200–1500) includes samples dating after the introduction of maize agriculture to the

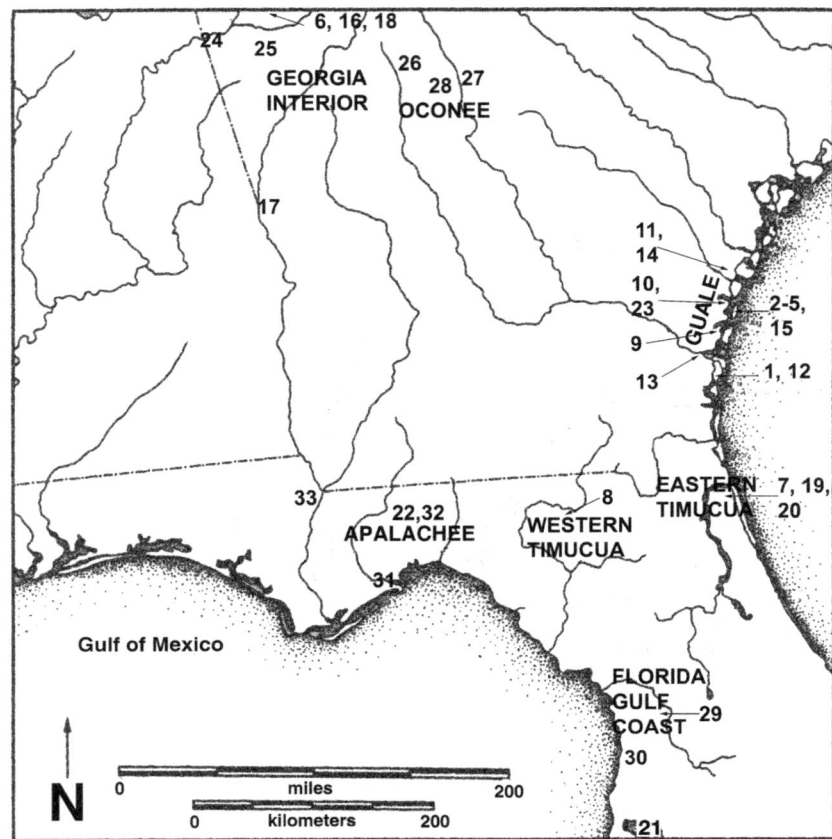

Figure 8.1. Locations of precontact and protohistoric period samples used in comparative analyses. Site names are associated with respective sample numbers in tables 8.1–8.3. Regional labels indicate sample aggregation strategy as determined by historical and archaeological data as well as natural gaps in the archaeological record of cemeteries.

southeastern United States but before contact, and the protohistoric period (A.D. 1500–1600) includes samples from burial contexts that are precontact in style (such as low mounds) but contain European-manufactured objects and, therefore, represent the mortuary activities of 16th-century populations before the transition to Christian interments.

Several geographic divisions that loosely correlate with hypothesized sociopolitical and linguistic boundaries are also defined for the purposes of comparative analysis. These divisions are, in part, also based on gaps that exist in the archaeological record. Regions in La Florida include Apalachee province, western Timucua (Utina), eastern Timucua (St. Johns River), and Guale province. In addition, samples from the Florida Peninsular Gulf Coast region, the Middle Oconee River region and the Georgia interior are included to provide a broader regional perspective. Although the location of the interior Georgia samples is not ideal, they do provide some estimate of biological variability in Georgia beyond the reach of the Spanish missionaries.

Samples from the early precontact period are listed in table 8.1. There are five samples from the Guale region, one sample from the Georgia interior, one sample from western Timucua (Utina province), and one sample from eastern Timucua. There are no early precontact Apalachee samples. No aggregation of samples is required for this analysis. Skeletal samples from the late precontact period are listed in table 8.2. There are seven samples from Guale province, three samples from the Georgia interior, two samples from eastern Timucua, one sample from the Florida Peninsular Gulf Coast, and one sample from Apalachee province. Skeletal samples from the protohistoric period are listed in table 8.3. There is one Guale province sample, two samples from the Georgia interior, three samples from the Middle Oconee River region, two samples from the Florida Peninsular Gulf Coast, and three samples from Apalachee province.

Table 8.1. Early precontact skeletal samples from La Florida and central Georgia

Site	Province	Position	Date (A.D.)	Reference
1. Airport (9Gn1)	Guale	St. Simons Island	ca. 500?	Larsen, 1982
2. Johns Md (9Li18)	Guale	St. Catherines Island	ca. 1100	Larsen, 1982
3. Marys Md (9Li20)	Guale	St. Catherines Island	ca. 1100	Larsen, 1982
4. McLeod Md (9Li47)	Guale	St. Catherines Island	ca. 100	Thomas and Larsen, 1979
5. Seaside Mds 1,2 (9Li26/62)	Guale	St. Catherines Island	ca. 500	Thomas and Larsen, 1979
6. Sixtoe Field (9Mu100)	Creek?	Georgia Interior	ca. 1000	Kelly et al., 1965
7. Mayport Md (8Du96)	E. Timucua	Florida Atlantic Coast	ca. 100	Wilson, 1965
8. McKeithen Md C. (8Co17)	Utina	Florida Inland	ca. 500	Milanich et al., 1984

Table 8.2. Late precontact skeletal samples from La Florida and central Georgia

Site	Province	Position	Date (A.D.)	Reference
9. Little Pine Island	Guale	Sapelo River/Coast	1200–1300	Larsen, 1982
10. Norman Md (9McI64)	Guale	Sapelo River/Coast	1200–1300	Larsen, 1982
11. 7 Mile Bend Md (9Bry6)	Guale	Ogeechee River/Coast	1200–1550	Larsen, 1982
12. Kent Md (9Gn51)	Guale	St. Simons Island	1300–1550	Larsen, 1982
13. Lewis Creek Mds (9McI88)	Guale	Altamaha River/Coast	1200–1300	Larsen, 1982
14. Irene Mound Site (9Ch1)	Guale	Savannah River/Coast	1300–1550	Hulse, 1941
15. South End Md (9Li3)	Guale	St. Catherines Island	1200–1550	Larsen, 2002
16. Little Egypt (9Mu102)	Creek?	Georgia Interior	1350–1550	Hally, 1979, 1980
17. Avery (9Tp64)	Creek?	Georgia Interior	1450–1550	Huscher, 1972
18. Bell Field (9Mu101)	Creek?	Georgia Interior	1250–1350	Kelly, 1970, 1972
19. Holy Spirit (8Du66)	E. Timucua	Florida Atlantic Coast	ca. 1500	Jones, 1991
20. Browne Md (8Du62)	E. Timucua	Florida Atlantic Coast	ca. 1200	Sears, 1959
21. Tierra Verde Md (8Pi51)	Tocobaga	Fl. West Coast	ca. 1350	Sears, 1967; Hutchinson, 1993
22. Lake Jackson (8Le1)	Apalachee	Florida Panhandle	1240–1475	Jones, 1982

Table 8.3. Protohistoric (postcontact, pre-mission) skeletal samples from La Florida and central Georgia

Site	Province	Position	Date (A.D.)	Reference
23. Pine Harbor (9Mc64)	Guale	Near St. Catherines	1425–1580	Cook, 1980; Larsen, 1990
24. King (9Fl5)	Creek?	Georgia Interior	ca. 1540	Blakely, 1988
25. Leake (9Br2)	Creek?	Georgia Interior	1500–1600	Wauchope, 1966
26. Dyar (9Ge5)	Creek?	Middle Oconee River	1520–1580	Smith, 1994
27. Joe Bell (9Mg28)	Creek?	Middle Oconee River	1580–1640	Williams, 1983
28. Woodlief (9Pm137)	Creek?	Middle Oconee River	1520–1580	UGA Accession
29. Tatham Md (8Ci203)	Tocobaga	Florida West Coast	1525–1550	Hutchinson, 1996
30. Weeki Wachee (8He12)	Tocobaga	Florida West Coast	1525–1550	Hutchinson and Mitchem, 1996; Mitchem, 1989
31. Snow Beach (8Wa52)	Apalachee	Florida Panhandle	ca. 1500	Magoon et al., 2001
32. Killearn Borrow Pit (8Le170)	Apalachee	Florida Panhandle	ca. 1500	Shapiro and McEwan, 1992
33. Waddells Mill Pond (8Ja65)	Apalachee/Chatot?	Florida Panhandle	ca. 1500	Gardner, 1966

Analytical Approach

The precontact database presents a number of challenges that preclude the use of standard multivariate statistical analysis (Kieser, 1990; Pietrusewsky, 2000) as well as quantitative genetic model-bound analyses such as those used in chapter 1 (Relethford, 2003; Relethford and Lees, 1982; Relethford et al., 1997). Sample sizes are small, preservation is poor, dental attrition is ubiquitous and often severe. These factors result in extensive amounts of missing data. The samples are also decidedly nonrandom. Sex estimates are impossible for many individuals (again due to poor skeletal preservation), and even more problematic, some samples cannot be divided into discrete individuals (Pine Harbor, for example). This last fact alone requires an analytical unit based on aggregate statistics above the scale of the individual. These challenges require use of a simpler approach based on the patterning of variable means with no regard for the range of phenotypic variability *within* each sample. This is obviously not an optimal strategy but is one that facilitates broad comparison of dental phenotypes while recognizing that small sample sizes may produce spurious estimates of within-sample variability.

To analyze patterns of biological variation, the aggregate-sex mean for each sample was calculated. This produces a square matrix of means with variables as columns and samples as rows. Sample means were then used as variables within a principal components analysis to extract factor loadings that represent aggregate measures of similarity based on the corpus of dental measurements available. This approach identifies commonalities among tooth sizes for different tooth classes and generates new variables which reflect common size and shape components. Each component is orthogonal and statistically independent, where the first principal component often represents a general size factor as indicated by equal and positive loadings for each variable. Samples that have uniformly large or small means for all variables are easily identified along this axis. Other components may represent vectors of tooth size proportionality—for example, mesiodistal versus buccolingual tooth size; maxillary versus mandibular tooth size; anterior versus posterior tooth size; or more specific configurations such as maxillary, anterior, buccolingual tooth size versus mandibular, posterior, mesiodistal tooth size. Using an aggregate approach such as this maximizes the number of samples and individuals that can be included in an analysis. However, the results are not directly comparable to those presented in chapter 1.

Sample-specific variable means are used to generate principal components for each time period. Bivariate plots are presented for the first two principal components with each axis scaled by the square root of the corresponding eigenvalue for that eigenvector. This places greater emphasis on components representing maximum variation. The pattern of affinity among samples is interpreted informally based on these "genetic distance" ordinations in reference to hypothesized social and political boundaries generated from early historic period documents on tribal and linguistic affiliations (see chapter 1).

Regional Comparisons by Time Period

Analysis of population structure for the early precontact period was performed using a subset of the original variable list (UI1BL, UCMD, UP1MD, UP1BL, UM1MD, UM1BL, LCMD, LP1MD, LP1BL, LM1MD, LM1BL, where UI1BL = Upper First Incisor Buccolingual, for example) due to missing data for some samples. Principal components analysis returned two factors with eigenvalues greater than one, representing 46 percent and 27 percent of the original variation in the dataset, respectively. A plot of the first two principal components is presented in figure 8.2 and several observations are noteworthy. First, the Timucua samples from the St. Johns River region (circles) are fairly distinct from the Guale samples (squares) and the single Georgia interior sample (diamond), while both Timucua samples are phenotypic nearest neighbors with each other. This suggests that the Timucua-speaking populations were more similar to each other, regardless of geographic distance, than to other populations. Second, there is no isolation-by-distance structure among Timucua and Guale province samples; Mayport Mound is more dissimilar from the Guale samples than McKeithen Mound C, despite the geographic proximity of Mayport Mound to the Georgia coast. Both of these observations suggest that the Timucua of the early precontact period were genetically isolated from their Muskogee neighbors to the north. Third, Sixtoe Field, from the Georgia interior, is clearly affiliated with the Georgia

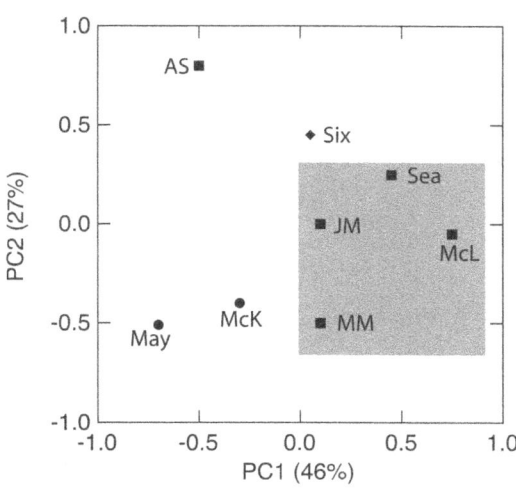

Figure 8.2. Principal components plot for early precontact period samples (ca. A.D. 500–1100). Gray square encompasses the Georgia coastal samples. AS = Airport site, JM = Johns Mound, May = Mayport Mound, McL = McLeod Mound, McK = McKeithen Mound C, MM = Marys Mound, Sea = Sea Mounds, Six = Sixtoe Field. Circle icon = Timucua samples, square icon = Georgia coast samples, diamond icon = Georgia interior sample.

coast Guale province samples to the exclusion of the Timucua samples. This suggests greater integration in the form of migration and gene flow between populations living along the Georgia coast and in the Georgia interior during the early precontact period (figure 8.3).

Analysis of population structure during the late precontact period excluded only two variables (UCMD, LCBL) due to missing data. Because of small sample sizes, several individual sites were aggregated within regions to implement the comparative analysis. Browne Mound and Holy Spirit were combined to form a single east Timucua sample, and Little Egypt, Avery,

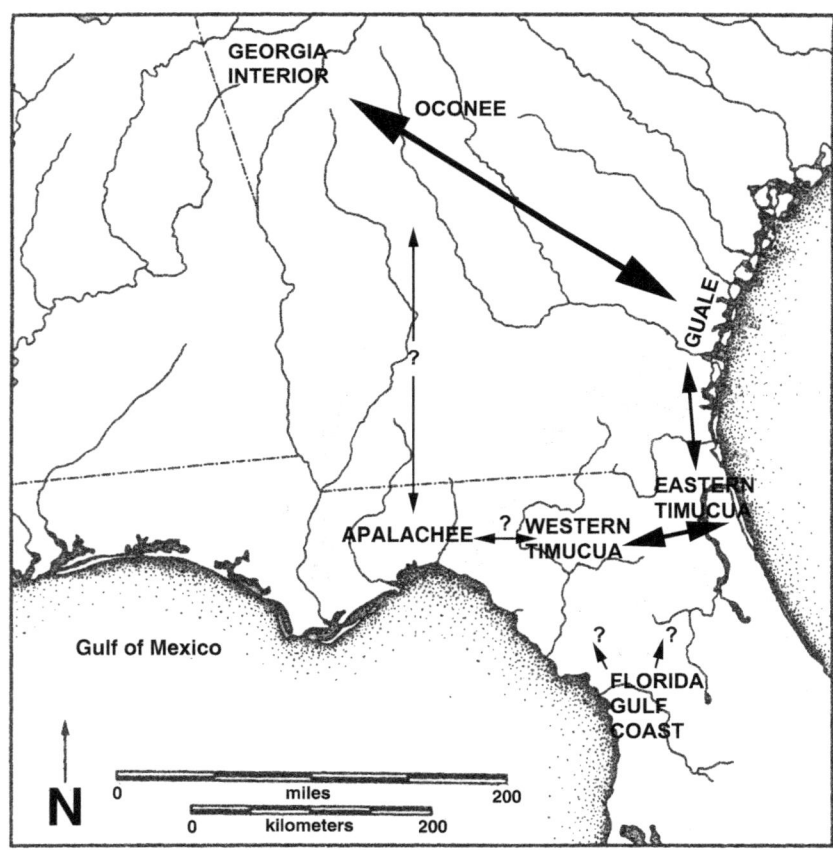

Figure 8.3. Degree of biological integration between different regions in Florida and Georgia during the early precontact period based on principal components analysis of tooth dimensions. The size of the arrows indicates the strength of integration between different areas. Question marks indicate a lack of data to estimate the degree of integration.

and Bell Field were combined to form an aggregate Georgia interior sample. Principal components analysis returned four eigenvalues greater than one, representing 34 percent, 22 percent, 16 percent, and 10 percent of the variation in the original dataset, respectively. The first component produced all positive factor loadings and represents a general size variable, whereas the second principal component represents the proportional difference in mesiodistal versus buccolingual dimensions.

A plot of the first two principal components is presented in figure 8.4 with a 95 percent confidence interval around the sample mean (the mean of means) for Georgia coast populations. Lake Jackson, the only Apalachee sample, is the most phenotypically distinct from the other samples and is most similar to the Tierra Verde sample from the Tampa Bay region. In fact, both samples are phenotypic nearest neighbors. Contrary to the early precontact distance ordination (figure 8.2), the eastern Timucua samples from the St. Johns River region are phenotypically similar to the Guale samples from the Georgia coast. In fact, the combined eastern Timucua sample is within the 95 percent confidence interval for Guale and is more similar to Georgia coast samples such as Kent Mound and Seven Mile Bend than either of these are to other samples from the Georgia coast. The Georgia interior sites of Little Egypt, Avery, and Bell Field are similar to the Georgia coast samples (though beyond the confidence limits) and are clearly affiliated with the Guale rather than with the Apalachee of the Florida panhandle. These data indicate that patterns of affinity changed slightly during the late precontact period (figure 8.5). Regional isolation-by-distance throughout Florida and Georgia, including Apalachee and the Gulf coast, reflected the effects of long-range mate exchange that occurred along two distinct principal axes: (1) from west to east, that is, from peninsular Florida (Apalachee) through west-central Florida and to the Atlantic coast (Guale and eastern Timucua), and (2) from the Georgia coast (Guale) toward the Georgia interior. There was a trend for increasing tooth size as one moves east from Apalachee (Lake Jackson) toward the Florida and Georgia coast. Along the Atlantic coast, St. Johns River valley populations (eastern Timucua) and those living along the Georgia coast (Guale) appear to represent a fairly integrated biological population. This same pattern of tooth size distribution was documented during the early precontact period and may be reflective of intertribal integration throughout the region, as also documented in chapter 1 using more rigorous F_{ST} statistics. Considering that Guale province is spatially equidistant from both the interior populations of Georgia and from Apalachee province, there was considerable biological interaction between the Georgia coast and Georgia interior and very little interaction between Apalachee and the Georgia interior. Because there was no Apalachee sample in the early precontact period analysis it is not possible to determine whether this distinction was also present during that time period. However, it is clear that the Georgia coast and interior were integrated in a way that persisted for several centuries, and that this degree of integration was not also evident for the Apalachee.

Analysis of population structure during the protohistoric period excluded four variables (UI1BL, LI2BL, LM1MD, LM1BL) due to missing data. As with the late precontact period, several sites were aggregated to implement comparative analysis. The samples from Apalachee province (Snow Beach, Killearn Borrow Pit, and Waddells Mill Pond) were too small to be considered separately and were combined to form a single aggregate sample. In addition, three sites from the Middle Oconee River region (Dyar, Joe Bell, and Woodlief) were combined to form one sample, as were the two sites from the west coast of Florida (Tatham Mound and Weeki Wachee). Principal component analysis returned two eigenvalues greater than one representing 71 percent and 12 percent of the variation in the original dataset, respectively. The first principal component displayed all positive factor loadings and represented general tooth size, whereas the second component represented differences in mandibular versus maxillary tooth size.

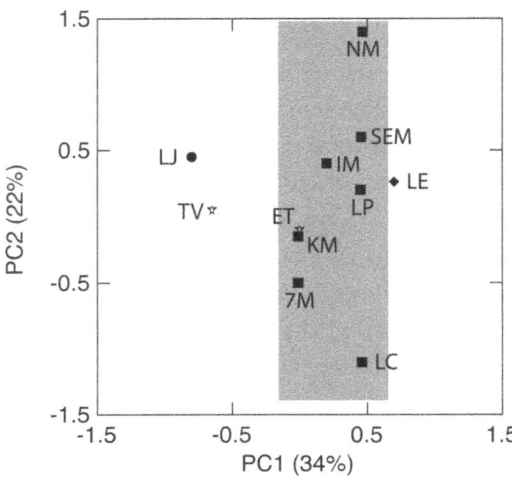

Figure 8.4. Principal components plot for late precontact period samples (ca. A.D. 1200–1500). Gray square encompasses the Georgia coastal samples. 7M = Seven Mile Bend; ET = Holy Spirit, Browne Mound; IM = Irene Mound; KM = Kent Mound; LC = Lewis Creek; LE = Little Egypt, Avery, Bell Field; LJ = Lake Jackson; LP = Little Pine Island; NM = Norman Mound; SEM = South End Mounds; TV = Tierra Verde. Square icon = Georgia coast samples, circle icon = Apalachee sample, star icon = Timucua samples, diamond icon = Georgia interior sample.

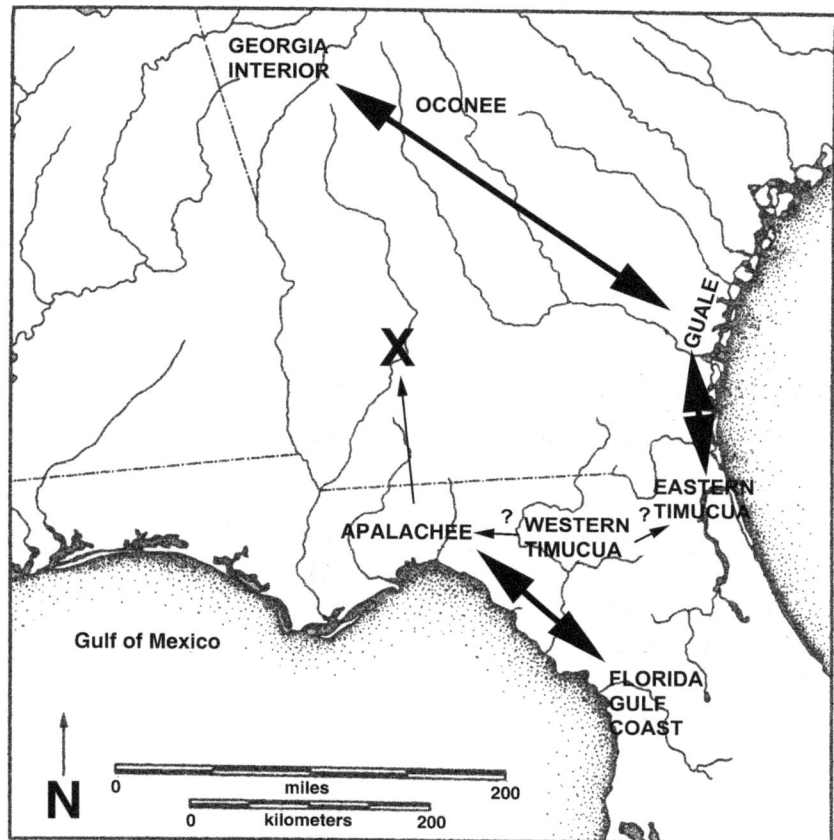

Figure 8.5. Degree of biological integration between different regions in Florida and Georgia during the late precontact period based on principal components analysis of tooth dimensions. The size of the arrows indicates the strength of integration between different areas. Question marks indicate a lack of data to estimate the degree of integration.

A plot of phenotypic distances between samples is presented in figure 8.6. There are several noteworthy observations. First, the King and Leake sites are so phenotypically similar and geographically proximate that they likely represent the same biological population. This may seem unsurprising but the similarity between these two samples supports the validity of the analytical approach used to generate the distance ordination. One would expect far less systematic patterning if the methodology used to represent patterns of phenotypic affinity was invalid. However, the most noteworthy observation is the complete change in regional population structure during

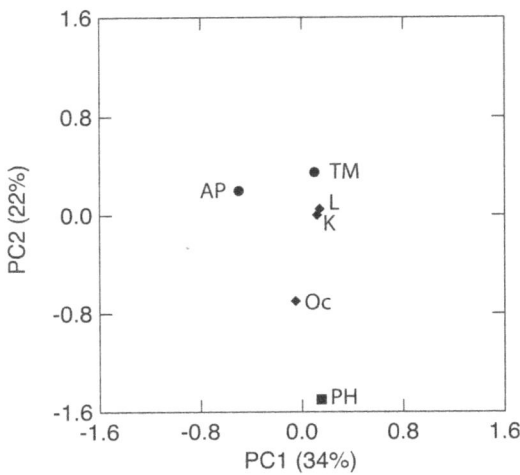

Figure 8.6. Principal components plot for protohistoric period samples (ca. A.D. 1500–1600). AP = Snow Beach, Killearn Borrow Pit, Waddells Mill Pond; K = King site, L = Leake site, Oc = Dyar, Joe Bell, Woodlief; PH = Pine Harbor; TM = Tatham Mound and Weeki Wachee. Square icon = Georgia coast sample, circle icon = Apalachee and Florida Gulf coast samples, diamond icon = Georgia interior samples.

the protohistoric period. Whereas the previous two phases witnessed biological integration between Georgia coastal and interior populations, during the protohistoric period the lone sample from Guale province (Pine Harbor) is now the outlier in the distance ordination. To the contrary, the Georgia interior populations are more biologically integrated with populations from western Florida, including those living in Apalachee province and farther south along the Gulf Coast of Florida (Tatham Mound and Weeki Wachee). The pattern of biological integration linking the Georgia coast and interior that had existed for approximately 1,000 years prior to European contact (from A.D. 500–A.D. 1500) was completely restructured during the protohistoric period (figure 8.7).

This pattern of biological variation is intriguing for several reasons. First, that this dramatic change occurred during the early 16th century suggests that the effects of European contact were manifest in broad-scale interaction patterns among indigenous communities. Whether this is due to the effects of epidemic disease or other disruptive processes is not knowable based on the biological data. This is, however, a potentially important new form of data that speaks to the New World pandemic debate. Second, the change in population structure is consistent with similar changes in ceramic technologies during the 16th century, and for some authors at least these changes have been interpreted as representing greater interaction between Apalachee and the Georgia interior (Mason, 2005: 182–185; Worth, 2002: 268–269). These data may, therefore, provide biological resolution to the debates about technological evolution. Clearly a biological signal indicates some degree of actual movement of people or genes was involved in the process and not just the exchange of ideas through diffusion. Third, the pattern of population relationships defines a new interaction dynamic in the early 16th century, and this dynamic would persist throughout the 17th and 18th centuries. The data, therefore, may belie changes in the biosocial landscape that preface the formation of the Creeks and then the Seminoles that followed, further supporting the ethnogenetic analyses presented in the preceding chapters.

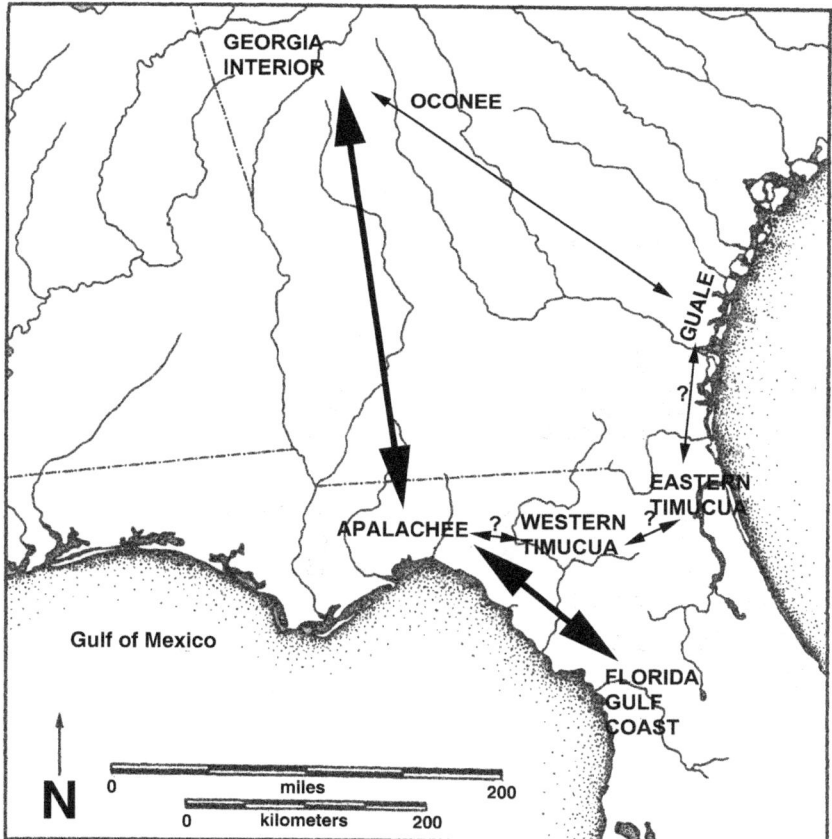

Figure 8.7. Degree of biological integration between different regions in Florida and Georgia during the protohistoric period based on principal components analysis of tooth dimensions. The size of the arrows indicates the strength of integration between different areas. Question marks indicate a lack of data to estimate the degree of integration.

The Myth of the Primitive Isolate

At the beginning of this chapter the question was posed whether the populations in La Florida were biologically distinct from those in what would become the Creek heartland. The answer to this question is very clear. At no time were the Georgia interior populations an outlier in the analyses, as one would expect if there was no gene flow or migration linking these communities together. In fact for about 1,000 years, from A.D. 500 to A.D. 1500, populations living along the Georgia coast and in the Georgia interior were biologically integrated to a significant extent in ways that contradict the baseline expectation of isolation-by-distance. In other words, the degree of integration was much greater between populations living 200 miles distant from one another (Georgia coast and interior) than between populations living within 50 miles of each other (Georgia coast and the St. Johns River eastern Timucua). This observation alone suggests that a significant element of sociopolitical identity is manifest in the phenotypic data; something was operating to prevent greater biological integration between populations in Guale and Timucua provinces. The pattern of population affinity changed during the protohistoric period, however. Integration between Georgia coastal and interior populations decreased as integration increased between the Georgia interior and communities located in western Florida—Apalachee and the western peninsular Gulf Coast in particular. Regardless of the time period considered, however, populations in the Georgia interior were never completely isolated from those living in what would become La Florida. And this contributes an additional aspect of biological history which affirms the hypothesized pre- and post-Spanish period biosocial continuity.

The first component was detailed in chapter 7. The earliest proto-Seminole communities to return to Florida were composed of remnant Christian Indians who had interbred widely with other ethnic groups during the 17th century. This connection provides a degree of direct biological ancestry between precontact populations of Florida and *some* of the earliest proto-Seminole bands. This interpretation is the most direct in that linkages are proposed between populations and identities proximate in space and time. However, it suffers from the tendency to view early Seminole history as too deterministic of later Seminole history. That is, even without the input and impact of pro-Spanish Christian Indians it is possible that the Seminole would have emerged anyway, although under different historical circumstances.

The second aspect of biological history that speaks to Seminole ethnogenesis is the record of Christian fugitivism during the 17th century. Because this process was occurring during the development of the Creek political body, some segment of the burgeoning Creek populace actually came from Florida. Therefore, the interaction dynamic during the 17th century negates any parsing of Florida and Georgia populations from a strict biological interpretive framework. That is, despite clear differences in social identities during the 17th century (see chapter 5), the transfer of personnel back and forth between La Florida and the Creek heartland directly positions the Seminole (as an offshoot of the Creeks) as biological descendants of the Florida tribes. Without contemporary 17th-century populations from the Georgia interior, however, it is not possible to evaluate the degree of migration or gene flow that linked these populations during the course of the formation of the Creeks. This remains a testable hypothesis for the future.

Finally, the rationale for dividing the Florida and Georgia groups, regardless of protohistoric ethnonyms or chiefdom affiliations, is further called into question by the analyses presented in this chapter. Going back as far as A.D. 500, at least, at no time was there a clear phenotypic distinction between populations living in La Florida proper and those living in the Georgia interior. In this sense, the fugitivism documented during the 17th century was a continuation of the flexibility of residence and population exchange which defined southeastern communities for over 1,000 years. These data suggest that the distinction between Florida's "ancients" and the Creeks and Seminoles is clearly an imposition of modern state borders on a continuous interaction and exchange network. The question of the Seminole as "true" Floridians becomes untenable as such.

Data presented in this chapter also contradict the primitive isolate view of human society in which monolithic social aggregates are seen as evolutionary units with permanence and temporal transcendence. This literature was discussed more fully in chapter 3, and it is important to stress at the close of this volume the lack of support for the "billiard ball" view of human history. However, from an operational perspective I too participated in the creation of social identities—that is, the imposition of these identities within a chrono-spatial framework. I also participated in the creation of biological populations through the strategy of sample aggregation for the purposes of implementing multivariate analysis with incomplete and fragmentary datasets. Nonetheless, the primary interpretive thrust of this monograph was to establish that the 17th century was a time of ethnogenetic transformation among Florida's indigenous communities living within the

Spanish mission system. One component of this process was the interaction dynamic between the mission communities and those in central Georgia. This same dynamic determined 18th-century historical trajectories (Creeks vs. Seminole) and also determined precontact interaction dynamics, at least as defined biologically.

And this is the point so well made by Patricia Wickman. While individual tribal or political identities may have waxed and waned during the 17th and 18th centuries, the fundamental cosmogony of the people was not destroyed by European influences, epidemic disease, or frontier violence and raiding. The *structure* of systems of integration may have changed through time, but the system itself remained as the true testament to indigenous history, as the biological data presented in this book so clearly demonstrate. As a continuing historical process, then, Seminole ethnogenesis can be pushed back even further—not in name but in the practices and social processes adopted by indigenous peoples as part of their negotiations with European colonials. As Wickman (1999) hypothesized, the patterns of biological integration may have changed through time, but the nature of the people did not. They may have been called Uchises, Yamassees, Apalachees, Timucuas, and Guales. But they were all Maskókîs, all the time.

Nine *Parting Comments*

The Bioarchaeology of Ethnogenesis

In exploring the relationship between the social and biological realms of the human experience I have attempted to place the study of human microevolution (biodistance analysis) within a broader field of social inquiry. I specifically argue that patterns of phenotypic variation are, in fact, heavily laden with identity symbolism. This isometry between biological and social identities results from the vast complexities of human reproductive systems, perhaps uniquely so, and the myriad social variables (religion, class, race, prejudice, socioeconomic status) that inform relations among individuals and among communities of individuals within a mate exchange network. In this sense, patterns of genetic variation can be "read" just like other indicators of social identities in the past—material culture, architecture, dress, and cuisine.

However, patterns of phenotypic variation, tooth size in particular, provide the added benefit of being invisible from those actors participating in the ethnogenetic process. That is, the social realm that flavors decisions about mate choice creates specific patterns of biological variation that we as anthropologists can observe. However, the actors involved in the discourse were unaware of the creation of these signals and therefore had little opportunity to manipulate them. Biology, governed by the rules of inheritance, cannot be manipulated or feigned and provides a direct representation of what people actually *did*. The body *is* material culture and is imprinted with indelible signatures of macroscale human behavior and group action at the population level.

In developing this relationship, this book ameliorates some of the tensions that exist within anthropology (and perhaps the social sciences in general) between the "biological" and the "cultural." The interdigitation of the two in the majority of scholarship seeks sociobiological and Darwinian explanations for social behavior, an approach many social anthropologists find unpalatable. This is certainly the case in the field of ethnic studies where one scholar in particular (Pierre van den Berghe) is the primary contributor, but not without much controversy. When a materials-based approach to the past is adopted, however, any disjuncture that may exist evaporates. By treating the body as a form of material culture, and by recognizing the intense social connotations underlying patterns of gene flow, the study of social process is significantly informed by *diachronic transformational* approaches to genetic variation. The argument is not sociobiological. The approach does not equate an ethnic group with a race or even with a breeding population, nor does it reify social phenomena within cladistic, taxonomic schemes. The goal is not to find the boundaries of an ethnic group and test whether it is coterminous with a breeding population. Finally, the approach does not consider ethnogenesis in a typological manner, that is, by trying to understand the origins of "a people" in terms of the mixing of distinct ancestral elements. Where possible, the data are left to speak for themselves and the emphasis is on reconstructing patterns of variation within a defined chrono-spatial framework, and then interpreting changes in the patterns of variation within a specific social, theoretical (ethnic identity and ethnogenesis theory), and historical framework.

Ultimately, the field will have to decide whether they agree with the approach adopted here. One could argue the data are over- or trans-interpreted; that is, based on one thing (biology) and interpreted in reference to another (social identity). However, there was never a firm statement as to middle-range linkages between mating behavior and ethnogenesis. This was intentional because I do not think the data in this book can speak to this issue, and in some ways it may not matter entirely. Nonetheless, the issue of whether gene flow is an active or passive component of social identity transformation remains a testable hypothesis. Does interbreeding between communities *cause* feelings of ethnic sentiment with an underlying basis in allele hegemony, as sociobiology would predict? Or does mate exchange cause feelings of social cohesion, communalism, and solidarity because of the universal bond between parent and offspring (itself Darwinian)? Perhaps the relationship is completely non-Darwinian and gene flow's importance rests in its ephemeral relationship to notions of kinship and the social obligations accorded to specific positions within these interpersonal relationships. Finally, we can ask whether interbreeding contributes nothing to feelings of ethnic solidarity and merely reflects, in a completely passive way, the complex nature of social interactions in the past. In other words, social, economic, and political factors stimulate ethnogenetic processes, and mate exchange is merely a reflection of these phenomena, a trailing indicator.

Colonial Ethnogenesis in the Southeastern United States

Using data on tooth size, a phenotype with a high degree of genetic determination, this book has examined how patterns of interpopulation variability and affinity changed through successive time periods among communities living in the Florida panhandle, north-central Florida, in and around St. Augustine and the St. Johns River drainage, and along the Georgia coast. This area experienced the greatest intensity of Franciscan proselytization during the 17th century. Data were collected for a series of precontact samples, four early mission period samples (ca. 1600–1650), and three late precontact samples (ca. 1650–1700), and it is important to note that the same local populations are approximately sampled for each time period. Odontometric data were analyzed using R-matrix methods, which produce estimates of interpopulation genetic distances as well as regional estimates of among-population genetic diversity. This analytical approach allows for varying estimates of narrow-sense heritability and has the ability to model different demographic scenarios by incorporating effective population sizes as input parameters. The approach focused on patterns of variation and affinity rather than group allocation, and the evolutionary interpretation of these data assumed a microevolutionary form. Gene flow, migration, and genetic drift were favored over mutation and selection; the latter two mechanisms of evolutionary change were inappropriate given the data type and short time periods considered.

I have chosen to interpret these data within a specific theoretical framework focusing on ethnogenetic theory and its predictions about intergroup biological and social integration. More specifically, I focused theoretically on the concept of ethnic identity and the mechanisms responsible for the formation and dissolution of ethnic identities when viewed in historical perspective. Operationally, the evolutionary analyses were embedded within existing scholarship that views the body as another form of material culture, and as such, can be interpreted in a way that informs the social realm. This was the primary theoretical advance offered by this book; practitioners of biodistance analysis rarely engage the broader social sciences and actively articulate with archaeological models from such disparate fields as stylistic analysis and practice scholarship. Gene flow does not have to be reduced to a passive matrix of personnel or genes moving across the landscape. Rather, when a more socially embedded theoretical framework is adopted, reconstructed patterns of gene flow can be reinterpreted as a relic of past social behavior and emotional attachment.

Results of the population genetic analyses are robust and allowed a relatively straightforward interpretation. At no time can it be said that the populations living within the Spanish sphere of influence were genetically heterogeneous. This may reflect data choice to some extent but also indicates that the inhabitants of Florida and Georgia were united by complex patterns of integration representing the movement of personnel and/or genes across long distances. This reflects the political structure of chiefdoms and their unstable positions within the larger sociopolitical system; collapse redistributed people across the landscape. At the same time, intense competition among rival chiefs necessitated incessant, small-scale warfare which often resulted in the capture of enemies (men, women, and children), and this had clear microevolutionary consequences. Communities were well and widely integrated, despite the sometimes extensive political buffer zones which physically separated chiefdoms. In this sense, it is noteworthy that the biological evidence supports what numerous archaeologists have inferred about the political structure of Mississippian chiefdoms in the Southeast. The general interpretation of the precontact period is one of biological homogeneity.

When these same populations are examined nearly a century later (during the early mission period), the structure of population relationships had not changed significantly; however, the degree of regional integration had declined. As measured statistically, F_{ST} nearly doubled between the precontact and early mission periods. What best explains this? From an evolutionary perspective it is almost certain that population size declined between the precontact and early mission periods, and this suggests genetic drift may have operated to reduce biological integration. However, previous analysis of these data (Stojanowski, 2005a) suggested that the a priori primacy of drift as an explanatory mechanism may not be appropriate. Evidence for epidemics is more limited and not geographically widespread. Bioarchaeological analysis of dental and skeletal pathologies does not suggest that morbidity increased immediately after contact across La Florida but rather assumed a more local and specific pattern. Not every church cemetery evinced images of massive epidemic mortality and overcrowding. And intrapopulation phenotypic variability did not universally decline from the precontact to the early mission period as predicted by evolutionary theory under a model of demographic collapse. For these reasons I propose that the transition to the early mission period also witnessed a decline in migration and gene flow among Catholic communities.

With respect to Hickerson's (1996) model of ethnogenetic change, the early mission period coincides well with the predictions of her separation phase in which the saliency of social identity categorizations declined. This is consistent with what is known about the nature of political integration and the warfare that went along with it. Specifically, it is known that those polities that received friars stopped fighting among themselves, and because warfare served an integrative function in prehistory, it follows that biological integration declined after contact. As with ethnic groups, political organizations in the precontact southeastern United States experienced their genesis, in part, through the very interaction that defined their existence. And, just as with ethnic groups, when the importance and durability of interactions declined, so did the importance of the social identities constructed in the context of that interaction sphere. Therefore, the initial stage of ethnogenesis in Florida witnessed a decline in genetic integration as a more localized form of discourse evolved. At the same time, when one withdraws from the classic position that epidemics preceded actual European contact, the early mission period is interpreted as a time of internal turmoil; large-scale changes in the nature of indigenous interaction patterns had yet to coalesce because communities still functioned at the local level. Migration was not widespread, allegiances were not strained, and marital and mate exchange proscriptions were not obviated by demographic collapse.

All of this changes during the latter half of the 17th century. Population genetic analyses clearly indicate that a single biological population was resident throughout La Florida, as represented by Apalachee, Guale, and perhaps Yamassee or Timucua samples that postdate the mid-17th century. Estimates of regional genetic diversity (F_{ST}) declined by 80 percent to the lowest levels observed. Population structure was nonexistent. Late mission period samples were genetically equidistant, suggesting that populations in La Florida were experiencing widespread mate exchange due to increased rates and lengths of migration. While in part reflective of the advancing rate of demographic collapse that may have necessitated long-range gene flow across ethnic boundaries, these data also reflect a more active role of indigenous communities in this process, which caused feelings of ethnic solidarity to coalesce.

However, the biological data do not exist in isolation; they do not provide the only evidence for ethnogenetic transformation. Consideration of previous historical and archaeological research provides abundant evidence that ethnogenesis was ongoing at the time that the missions were destroyed, or at least that the stimuli were present for ethnogenesis to occur among indigenous Christian communities. Much of this interpretation hinges on the mid-century date—the transition to the liminal phase occurred around 1650. There were objectifying elements which externally defined the experiences of Florida's faithful as well as subjective or emic processes that corroborate the biological inferences.

For example, the Spanish treated all mission Indians, regardless of language or culture, in a fairly similar manner and actively ignored nuances of ethnic or cultural diversity. This is best reflected in Spanish policies of aggregation and labor taxation which reified the division between Indian and Spaniard, even blurring the lines of faith by allowing pagan populations to settle among the converted. Dealings with the colonial government were, in other words, one dimensional and accorded no special status to distinct ethnic groups. The timing of ethnogenetic change, however, has more to do with hemispheric politics. When the English began to expand their trade interests into the Southeast (emanating initially from Jamestown), conflict among native communities escalated as they were positioned between Spanish and English interests geographically and symbolically. Florida Catholics as vassals of the Spanish Crown were embroiled in the struggle for control of the New World, yet they were poorly armed with inferior weapons, and this created a dramatic power imbalance among native communities which further refined ethnic boundaries. Slave raiding intensified after 1650, and this horrified the converts in La Florida, catalyzing patterns of antagonistic opposition. English-allied or English-serving, proto-Creek, opportunistic and allochthonous raiders, and Spanish loyalist identities existed within a large tribal zone that encompassed much of the southern half of eastern North America, with input factors from as far away as the Great Lakes and output effects reaching into West Africa. The World System was indeed in effect, and a myopic approach that fails to consider local and global histories is simply not sufficient.

None of these objectifying factors by themselves guarantee, or even predict, the florescence of social cohesion among communities dispersed over hundreds of square kilometers. However, increasing rates of migration and fugitivism brought together peoples of various backgrounds in polyethnic settings like St. Augustine or mission San Luis. It was in contexts like these where information exchange engendered feelings of solidarity, racial inferiority (the Spanish "Republics"), or social invisibility (treatment like chattel). Life course expectations and experiences had become quite similar, and this caused feelings of ethnic sentiment to develop regardless of antagonisms shared by preceding generations. All such rivalries and conflicts lost meaning in this new world. Language, dress, and posture; diet; health experiences and the shared misery of periodic epidemics; ideological conversion and changes in funerary rituals; the fear of being captured and sold as a slave; conspiring with the Spanish in reprisals against the English; and changes in the built environment and the material world fostered shared ethnic sentiments. These factors, in a highly objectifying world, caused ethnogenesis, and did so in a manner consistent with predictions of Hickerson's (1996) life cycle model of ethnogenetic change.

The long view of history does not define "periods" for the purpose of compartmentalizing continuous processes, and social histories are no exception. The "story" of Florida's precontact indigenous tribes does not end in 1706 (or 1763), but rather the social processes implied throughout this book set the stage for the continuation of the ethnogenetic process which began with first contact in the 16th century. Historians of Seminole history, for example, highlight the distinct social lives and political experiences of those "Creeks" living in Georgia as opposed to those in Florida. Over time, these differences resulted in Seminole ethnogenesis, a fission not a merger (as was occurring during the 17th century), that resulted in the distinct tribal identities still in existence. However, if spatial distance really contributed to burgeoning social distance, then the ultimate explanation for Seminole ethnogenesis must address the reasons the earliest proto-Seminole populations migrated to Florida during the first quarter of the 18th century. Various push (souring English relations) and pull (Spanish invitations and fertile land) mechanisms have been proposed to explain this migration from central Georgia. However, the widespread genetic homogenization documented by the analyses presented in this book, combined with the documented rate of out-migration during the 17th century, the Apalachee exodus after Moore's assault on the missions in the early 18th century, and the vagaries of colonial ethnonomy all suggest a more continuous narrative of Floridian biosocial history—one that explicitly positions the Florida Seminole as direct biological descendants of Florida's precontact populations. That the earliest proto-Seminole populations to migrate to Florida settled in the Apalachee district is too coincidental considering that the leaders of these bands were married to Apalachee women, were themselves Christian in many cases, and were staunch supporters of the Spanish in their struggle with the English. Seminole ethnogenesis, therefore, resulted from a homecoming of pro-Spanish, Christian sympathizers and was not an invasion by a nonlocal, postcolonial ethnic amalgamation with no connection to the past.

Nonetheless, this model of Seminole ethnogenesis which links the Spanish period ethnic identities to the 18th-century Seminole relies on a very specific set of historical circumstances and processes. In addition, it evokes a linear and somewhat biological view of social identities by assuming an a priori distinction between the proto-Creeks in Georgia and the "true" Florida Indians who would receive the friars in the 16th and 17th centuries. However, when additional data are added from samples from the Georgia interior, it is evident that these populations, divided by several hundred miles, were biologically integrated to a great extent. In particular from about A.D. 500 to A.D. 1500, populations living in the Georgia interior were nearly identical phenotypically to those living along the Georgia coast, but not in western Florida. The division between Georgia and La Florida Indians is, therefore, completely illusory from the perspective of evolutionary biology. However, during the protohistoric period, immediately after contact but prior to the establishment of permanent missions in La Florida, the interaction dynamic changed significantly. In particular, Georgia interior populations were biologically integrated with populations living in western Florida, and the Georgia coast was now somewhat isolated from the remainder of the region. This may reflect the effects of epidemics or simply sociopolitical disruptions caused by 16th-century entradas and colonization efforts. Regardless of the explanation for the change in interaction dynamic, however, it is interesting that exactly the same north-south migration pattern defined 17th-century interactions between Florida and Georgia indigenous populations. And furthermore, this structure continued to define migration patterns in the 18th century, when groups of disaffected "Creeks" started to be called "Seminoles" when living in Florida.

Bioarchaeology, as a fairly new discipline, can contribute uniquely to our understanding of social processes in past populations. And it can do so in ways that complement historical, archaeological, and anthropological genetic approaches, producing unique inferences that cannot be gleaned from other sources. Here I have emphasized the microevolutionary component of bioarchaeology and stressed that biodistance analyses based on skeletal phenotypes can produce truly remarkable observations about past mating behavior that are simply not possible using modern genetic techniques and would never be written in the historical annals. The time depth available through archaeological sampling is the critical component of success and in my opinion completely makes up for the problematic aspects of polygenic, multifactorial inheritance (namely, genotype-environment interaction). As a marker of identity, dental phenotypes are particularly interesting because of their invisibility to those participants in the process of identity transformation. As such, these data represent the actions of groups of people which I argue reflect the complex identity discourses that occurred in the past. Yet these people had no knowledge of the material record they were creating of these actions, thoughts, and dispositions. Therefore, biodistance analyses nicely complement more visible elements of biosocial research such as cranial and dental modification, topics that represent a large portion of scholarship that links the material skeletal record to past social identities (see Knudson and Stojanowski, 2009).

It is also now common to stress the importance of context as bioarchaeology moves away from purely descriptive approaches (Buikstra and Beck, 2006). Given the structure of this work, with the analytical results presented in bookend fashion, it is clear that the interpretation of bioarchaeological signatures greatly benefits from deep embedding within an appropriate historical context. In fact, the primary purpose of this book, as opposed to my previous writings on this topic (e.g., Stojanowski, 2005c), was the further development of the contextual basis of interpretation. Such leeway is not available in most journal formats, and the expansion of bioarchaeological titles from the presses reflects this.

Likewise, it is common to focus bioarchaeological research on problems that engage the broader social sciences and utilize social theory. It is less common, in my reading of the literature, to produce work that directly advances our understanding of issues of contemporary relevance. The approach outlined in this book has this potential but does not itself achieve this goal. Identity centers much of the conflict that exists in the world today, yet we are so very far from understanding how these social forms develop in a historical perspective. Determining what role, if any, marriage and mate exchange have in this process is crucial, and bioarchaeological approaches with their diachronic frameworks have the potential to offer novel perspectives. These types of social issues with contemporary relevance are the kinds of topics that bioarcheologists must engage to remain relevant to the world, and as recent emphasis on social identity suggests, the field is actively doing so. I hope that the ideas set forth here will inspire future attempts that achieve even greater clarity and sophistication.

Notes

Introduction. The Bioarchaeology of Ethnogenesis

Epigraph notes. Mucius Scaevola is a legendary Roman hero who thrust his right hand into a fire in disdain for his Etruscan adversaries. He has since been promoted as an icon of staunch resistance. Rodrigo Ranjel was Hernando de Soto's private secretary. His account of De Soto's expedition, reproduced in Oviedo's mid-16th-century text, *Historia Generale y Natural de las Indias*, was translated and edited by Yale historian Edward Gaylor Bourne in 1904.

1. I use the term *mechanical* to refer to evolutionary processes defined by mathematical population genetic models that do not incorporate aspects of the social context in which mate exchange occurs. In the case of demographic collapse, genetic drift is expected to change gene frequencies as population size declines.

2. It is important to differentiate racial science versus racist science. Racist science assumes a hierarchy of human populations with explicit differences in intelligence and capacity related to these vertical schemes. Racial science adopts an outmoded, typological view of human variation and is not racist per se. Racial views of the human past reify human populations into discrete entities and approach human diversity in terms of a limited number of essential types while ignoring variation among individuals within those types.

3. See Bowden and Goose (1969), Corruccini and Potter (1980), Dempsey and Townsend (2001), Dempsey et al. (1995), Goose (1971), Harris and Smith (1980, 1982), Harzer (1995), Hu et al. (1991), Kolakowski and Bailit (1981), Potter et al. (1983), Sharma et al. (1985), Townsend and Brown (1978a, b, 1979), Townsend et al. (1986, 2003) as compared with Hauser and DeStefano, (1989), Scott and Turner (1997), Sjøvold (1984), Sparks and Jantz (2002), Susanne (1977).

4. The polar teeth are for the maxilla: the first incisor, canine, first premolar and first molar. For the mandible the polar teeth are the second incisor, canine, first premolar, and first molar. The concept of polar teeth is still used in bioarchaeology and dental anthropology but is not supported by modern analyses of tooth genetics.

Chapter 1. Genetic Landscapes of Spanish Colonial Florida

1. Whitehead (1992: 139) notes that 1652 was the date the Spanish Crown officially ended the *conquista a fuego y sangre* in favor of the *conquista de almas*. Pedro Menéndez de Avilés, in a letter to a Jesuit Friend dated 1566 noted, "it is wasted time to think that the Holy Gospel can be established in this land with the army alone" (Menéndez letter to a Jesuit Friend, in Bennett 1968: 157).

2. Note, however, that Milanich (1999: 42) diverges from Worth (2004) in his assessment of Guale political structure, noting, "[they] were never united as a single political unit." Milanich (1999) included Orista within the Guale chiefdom.

3. The territorial limits of the Guale polity are subject to considerable debate. Larson (1980) notes that most Guale towns were located within 15 miles of the coast. However, Worth (2004) places the Guale village of Tulafina as much as 25 miles from the coast and notes the difficulty in establishing a political affiliation for this village and thus establishing a western boundary for the Guale chiefdom. A 1597 journey to the Georgia interior was recorded as follows: "after a trip of eight days, seven of which were spent in jungle country, the explorers arrived in the town of Tama" (Lanning, 1935: 112). Although the southern boundary of the Altamaha River does correspond with the distribution of prehistoric Irene phase ceramics (Saunders, 2000b), the northern boundary of Guale, here presented as around the Ogeechee River based on Worth (2004), has been debated. The issue is whether the chiefdoms of the Port Royal area (Orista and Escamaçu) were included within the Guale paramount chiefdom. Jones (1978, 1980) supports this opinion whereas Worth (2004) disagrees. Bushnell (1994) pushed the northern Guale boundary to Edisto Island. Hann (1987) places the northern boundary at the Savannah River, thus excluding the Port Royal chiefdoms but pushing the boundary north of the Ogeechee. Hann (1987) and Swanton (1946) align the Port Royal groups with those in the Georgia interior.

4. Laudionnière reported, "during the winter they retire for three or four months into the woods where they make little huts of palm leaves and live on acorns, fish . . . oysters, deer, turkeys" (Bennett, 1975: 15). When left without provisions, Laudionnière commented further, "the Indians are accustomed to leave their houses and retire into the woods for a space of three months, namely January, February, and March" (Bennett, 1975: 121).

5. Milanich (1999: 41) notes, "It was long thought that the Guale spoke a Muskhogean language, a language closely related to one spoken by Creek Indian ancestors who lived westward in interior Georgia. But recent research by anthropologist William C. Sturtevant has questioned that notion, and the question of the linguistic affiliation of Guale is unresolved." Worth (2004: 238) notes that Domingo Agustín Báez compiled linguistic data on the Guale language but that the manuscript has not been found. See also Swanton (1922: 85).

6. Hann (1996a) discusses the history of the use of the term *Timucua* to refer to speakers of this language group, noting that it is unknown what term, if any,

Timucua speakers used themselves to denote their shared linguistic affiliation. In Laudonnière's account (Bennett, 1975) it is evident that *Timucua* was used to refer to the enemies of the local St. Johns River area populations.

7. The names and locations of specific Timucua chiefdoms have changed over the years. Goggin (1953) produced a four-part geographical model: east = Freshwater, Saturiwa, Tacatacuru, Yui, Icafui, and Yufera; west = Potano, Timucua, Onatheaqua, Yustega; southeast = Acuera, Ocale; southern = Tocobaga, Ucito, Pohoy, Mococo. More recent scholars have completely discounted the southern group as being Timucua. Deagan (1978a) listed the following associated with the eastern Timucua: Cascange (subsuming Icafui), Yufera, Yui/Ibi, Saturiwa, Agua Dulce, Acuera, Tacatacuru. Milanich (1978) specifies west (Potano, Ocale, Utina, Yustega) and east (Saturiwa, Acuera, Freshwater, Cascange, Icafui, Tucururu, Yui, Yufera, Tacatacuru) groups. Interestingly, despite concurrent publication in the same volume Deagan and Milanich parse the eastern groups differently: Deagan combines Cascange and Icafui and omits the Tucururu. Milanich (1996) provides the familiar list of 11 (Ibi, Icafui, Oconi, Yufera, Tacatacuru, Saturiwa, Utina, Yustaga, Potano, Acuera, Ocale) and adds an additional 14 tribal affiliations (Caravay, Alicamani, Omoloa, Malica, Casti, Seloy, Coya, Chilili, Calanay, Edelano, Enecape, Molona, Patica, Arapaha). With the exception of the last of these, these tribal names are associated with the Atlantic coastal and southern interior regions, which were quickly devastated by early slave raids and epidemic disease. Milanich (2000) changes Utina to Uriutina and drops Ibi; 24 additional chiefdoms are listed (Tucuru, Astina, Chilili, Coya, Edelano, Enecape, Molona, Omittaguq, Onachaquara, Outina, Patica, Seloy, Ibihica, Alimacani, Caravy, Casti, Guadalquini, Malica, Napa, Omoloa, Aguacaleyquen, Cholupaha, Napituca, Arapaha plus the Mocama area groups). This list remains largely unchanged in Milanich (2004). The Tucuru are added as speakers of Acuera and a more general term *Indians of La Costa* is added.

8. Hann (1996a: 7) notes, "the vocabulary differences among the dialects that Pareja chose to record appear to be relatively few. In his grammar he noted only eight for Timucua Province's dialect, four for Potano, three for Yufera, two for Agua Salada, and one for Ocone."

9. The broader affiliation of Timucua languages has been much debated with Muskhogean, Siouan, and Arawakan affinities proposed. Granberry (1993) has found compelling morpheme and lexeme similarities between Timucua languages and those of the Orinoco Delta in Venezuela, most specifically Warao. Milanich (2004) finds no archaeological evidence to support a migration of South Americans to Florida, however.

10. Lanning (1935: 11) notes that when Governor Ybarra visited Guale and Timucua in 1604 he changed interpreters as he passed from one province to the next: Juan de Junco in Guale and an Indian named Santiago in Timucua.

11. Effective population size differs from census population size. The latter is the total number of individuals living within a population, the colloquial use of population size. However, from an evolutionary perspective, individuals who do not directly contribute to the genetic variability transmissible to the next generation are

not relevant. Such nonreproductive individuals include those who have not reached puberty as well as postreproductive members of a population. As a general rule the effective population size for human populations is estimated to be one-third to one-half the census size, although considerable variability exists because of differential sex ratios, differences in the age structure of the population, population structure, and marital practices. Hawks (2008) provides a readable discussion of the concept of effective population size and its importance in human evolution.

12. San Miguel de Asile is of questionable affiliation and may represent either a Yustaga or Apalachee mission community (Larsen, 1993).

13. Intra-observer error results from differences in the manner in which one observer records data from day to day. Generally, measurement acumen improves throughout a project such that the earliest data collected are the least precise and accurate. Intra-observer error also results from differences in lighting conditions, caliper calibration, and the general environment in which data are recorded. Inter-observer error occurs when different observers record data from different samples and those samples are then combined in an analysis. If the observers use different calipers or interpreted the measurement definitions differently then there may be systematic bias in the dataset.

14. See also Alvesalo and Tigerstedt (1974), Bowden and Goose (1969), Corruccini and Potter (1980), Dempsey and Townsend (2001), Dempsey et al. (1995), Goose (1971), Harris and Smith (1980, 1982), Harzer (1995), Hu et al. (1991), Kolakowski and Bailit (1981), Potter et al. (1983), Sharma et al. (1985), Townsend and Brown (1978a, b), and Townsend et al. (1986, 2003).

Chapter 3. Ethnogenesis, Social Identity, and Human Biology: A Bridging Model

1. As a simple but effective bibliometric support for this statement, searches for "ethnicity" and "ethnogenesis" in Web of Science returned 24,180 and 212 references, respectively.

2. Arutiunov (1994: 81) differentiates ethogony from ethnogenesis. The former refers to the evolution of the first ethnic groups from "amorphous early Paleolithic states devoid of ethnicity" and ethnogenesis refers to the fission and fusion processes that occurred after this initial formation.

3. William Sturtevant is credited with popularizing ethnogenesis in American social anthropology. Moore (1994a) attributed Sturtevant's inspiration to his travels to the Soviet Union in the 1960s where the concept had been widely used by linguists, social anthropologists, and race scientists to explain tribal formations of Central Asia (see Arutiunov, 1994; Bromley, 1974; Dragadze, 1980; Lindstrom, 2001; Slezkine, 1996, for further discussion of Soviet ethnogenetics). However, historical precedence for use of this term in anthropology lies with European biological anthropologists, although in a completely different form from that recognized today (Kozintsev, 1992; Michael, 1962; Oshanin, 1964; Wierciński and Bielicki, 1962).

4. Given no consensus on the definition of ethnicity or on the definition of "community," it is easy to understand how difficult it is to differentiate these two ephemeral social categories (Jenkins, 1997). Turner (1969: 126) notes that as early as 1955 definitions of community had proliferated beyond management, and the same can certainly be said of ethnicity (Banks, 1996; Jones, 2002). Both ethnic and community identity manifest during the process of interaction, intensify when threatened, and are largely considered situational and imagined constructs. Recent archaeological considerations have defined these terms as follows. A community is an "ever-emergent social institution that generates and is generated by supra-household interactions that are structured and synchronized by a set of places within a particular span of time" (Yaeger and Canuto, 2000: 5). An ethnic group is a "group of people who set themselves apart and/or are set apart by others with whom they interact or co-exist on the basis of their perceptions of cultural differentiation and/or common descent" (Jones, 2002: xiii).

Clarity remains elusive. Territoriality may be one defining component of a community; however, this is rejected in most formulations (Cohen, 1985; Turner, 1969; Yaeger and Canuto, 2000, but see Isbell, 2000). Cohen (1985: 104) distinguishes ethnicity from community on the basis of endurance permanence, that ethnicity is somehow related to disadvantage and subordination and is therefore less ephemeral. In fact, most definitions of ethnicity make some reference to fictive kinship or belief in descent from a mythological ancestor. To Yaeger and Canuto (2000), the distinction between ethnic and community identity is based on the frequency of interaction. Communities require frequent interaction for reproduction while ethnicities seem to have historical momentum due to "'imagined' essential characteristics" (Yaeger and Canuto, 2000: 6).

I personally see two distinguishing criteria. The first is the importance of ethnonymic reification for ethnicity to persist, which is, in practice, often another form of etic ascription. Communities "are." Ethnicities "are" because of a distinct referential term which is more clearly symbolic, and it is this factor that increases historical connections beyond the here and now. Second, while both community and ethnic identities develop in the context of interaction, in practice the interaction that defines a sense of community seems more internalized in orientation while interaction in the sense of ethnicity seems more externalized and differentiating. This reflects an emic view in each case and highlights Turner's (1969) distinction between *communitas* and Durkheim's solidarity.

5. Interestingly, Banks (1996) and Cohen (1969) consider Barth a primordialist, although, as noted by Eriksen (2002: 54) he is usually considered an instrumentalist. See also Jenkins (1997: 45) on this matter.

6. Ethnogenesis is not just seen as a non-Western phenomenon, as many cases of European ethnic and nationalist emergence indicate (Buncak and Piscova, 2000; Chropovsky, 1982; Hoppenbrouwers, 2006; Kaufmann, 1999; Kuzeev and Mukhamediarov, 1992; Lund, 2005; Mitchell, 2005).

Chapter 4. From Tribe to Ethnic Group, or from Ethnic Group to Tribe: Ethnic Demobilization during the Early 17th Century

1. Fontaneda fancifully suggests that integration may have occurred even across the Straits of Florida: "Anciently, many Indians from Cuba entered the ports of the Province of Carlos in search of it [the River Jordan]; and the father of King Carlos, whose name was Senquene, stopped those persons, and made a settlement of them, the descendants of whom remain to this day" (d'Escalante Fontaneda in True, 1945: 29). See Hann (2003: 27) for discussion of the deficiencies with the translation in the True (1945) edition. However, similar contact between Florida and Caribbean populations is suggested in Herrera's account of the de León expedition. He noted that the Lucayos Indians (Bahamas) referred to Florida as Cautio (Davis, 1935).

2. The use of existing trails by de Soto was so well established that Hudson et al. (1984: 66) state, "this has been accepted by researchers for many years and needs no further elaboration."

3. In some cases it is unclear whether Fontaneda was reporting information secondhand.

4. Both examples of the apparent freedom of movement and integration with local indigenous communities contrasts with Biedma's (in Bourne, 1922: 4) description of Juan Ortiz (a Narváez survivor) who could "tell us of nothing twenty leagues off."

5. As Reilly (1981) notes, the exchange between Carlos and Menéndez related to internal political strife within the Calusa chiefdom and did not directly involve regional political affairs.

6. Henige (1997: 166) outlines the discrepancies between the accounts of Ranjel, Elvas, and Garcilaso de la Vega about the specific nature of the women provided to de Soto by Casqui and Pacaha.

7. Hann (1988) discusses the relationship between the Apalachee and their Timucua neighbors as one of constant tension manifest in raiding and warfare. For example, in 1528 the Narváez expedition reported warfare between Apalachee and their Timucua neighbors (Núñez Cabeza de Vaca in Bandelier, 1905: 22). During the de Soto entrada, Ranjel (in Bourne, 1922: 73) reported, "seven chiefs from the vicinity came together, and sent to say to the Governor that they were subjects of Uçachile [Yustaga/Yustega], and that by his order and of their own will, they wished to be friends of the Christians and to help them against Apalache, a mighty province hostile to Uçachile and to themselves." Conflict between Apalachee and western Timucua chiefdoms persisted until the early 17th century when Fray Martín Prieto negotiated peace among these groups in 1608 (Oré 1936: 114–117). This peace was long-lived and initiated a period of increased communication throughout northern Florida (Hann, 1988: 183; 1996: 167). In 1596 Father Marrón commented on the peace which existed between the Timucua, Guale, and Agua Dulce; he credited this to the presence of the missionaries (Geiger, 1937: 67).

Chapter 5. The Liminal Phase of Ethnogenesis:
Objectification within the Eastern Woodlands Tribal Zone

1. As noted by Bolton and Ross (1925: 8–9), "The chief of Santa Catalina (St. Catherines) Island, where Menéndez landed, was an old man called Guale. From this circumstance the name Guale was specifically applied to the island; but it was gradually extended down the entire Georgia coast islands and mainland from Santa Elena to the St. Johns." See also a discussion of the origin of Guale and "Ouade" in Wallace (1975: 170).

2. Despite the fact that Spanish La Florida and St. Augustine were established in direct reaction to mid-16th-century French machinations at Charlesfort and Fort Caroline, the French exerted more limited influence on the colony for much of Spain's tenure in Florida. By the end of the 17th century France started to seriously look beyond Canada in its interests, and this too reflected a response to the success of the English along the eastern sea board. Particularly striking was the attack on the La Chua ranch located deep in the Florida interior in 1682 and again in 1684 (Bushnell, 1978: 428–429). Pirates had been reported near the St. Marks River (the port of Apalachee-Havana trade) in 1677 and destroyed the fort built here in 1682. La Salle attempted a colony along the Gulf Coast in 1685 which prompted Spanish reinterest west of Apalachee (Leonard, 1936: 547); however, this was already late in the mission period. Hann (1988: 323) reports of a 1686 expedition sent from Apalachee to locate the French in the Mississippi Valley. The posts at Biloxi (1699), Mobile (1702), and New Orleans (1718) were established when Spanish Florida was clearly in its twilight.

3. It is incorrect to assume, however, that prior to 1650, England and Spain had completely amicable relations. Sir Francis Drake, for example, sacked St. Augustine in 1586 and pirating activities by multiple European nations dogged the Caribbean throughout the 17th century.

4. The lack of plantations is why the African population of Spanish Florida remained comparatively small (Hoffman, 1997; Landers, 1997). Although slaves were brought with the earliest colonization efforts (Bolton, 1921: 141; Gannon, 1992: 321; Lyon, 1974: 414) and were a part of the early 16th-century entradas (Landers, 1997), they were never a significant presence in Spanish Florida or colonial St. Augustine. The proportion of Africans in St. Augustine increases only after the founding of Charles Town and the residual population flow of escapees from the English settlements farther north.

5. There is substantial evidence for early 16th-century slave raiding along the Florida coast. In a note in the Fontaneda memoir it is recorded, "Since the year 1510 . . . flotas and vessels have gone from these kingdoms to occupy Florida" (True, 1945: 61). Ponce de León mentioned this as one reason for his travels to Florida (Davis, 1991: 42), and Sturtevant (1962: 46) suggests that he was preempted by some unnamed and unsanctioned slave raids emanating from the Caribbean. Buckingham Smith reports slave raiding near Santa Elena in 1520 by "two caravels from Española"(True, 1945: 56). Pedro de Quexos and Francisco Goudillo carried

out slave raids in 1521 and 1522 (Priestley, 1928: 20). Lucas Vázquez de Ayllón organized slave raids along the Atlantic coast in 1521 (Jones, 1980: 217; Saunders, 1992: 140) and the Fidalgo of Elvas mentions that Vasco Porcallo de Figueroa went to Florida for slaves for his plantation (Elvas in Bourne, 1922: 34). The Italian explorer Verrazano visited Florida in 1524 and stole a woman's child "to bring into France" (Hakluyt, 1810: 360).

6. In 1662 Virginia actually outlawed the fur trade with groups farther north (Rountree, 1993: 74).

7. Gallay (2002: 296) estimates that at least 20,000 Florida Indians had been captured as slaves by about 1715.

8. Barbados was settled in 1637 and, like Jamestown, remained a problematic endeavor for its first few decades. Tobacco and cotton farming were failures. However, sugarcane was introduced from Brazil in the 1740s and a period of fevered exploitation commenced (Edgar, 1998: 26–28). By 1655 Robert Venables and William Penn had seized Jamaica from Spanish control.

9. As an ultimate demonstration of the world system perspective (Wolf, 1982), West African tribal conflict and slave raiding intensified concurrently, circa 1650 (Law, 1992).

10. Another reason that the English preferred African slaves over Native American slaves was that the latter often died of epidemics shortly after capture and also were able to escape and rejoin their people who could be incited to revolt. It was preferable to sell North American slaves to Caribbean plantations for this reason because there was no possibility of escape (Perdue and Green, 2001: 60).

11. Hann (1996a: 269) considers the Chiluque of the 1680 raid to be ethnically Yamassee as well.

12. Hann (1988: 185–186) provides an interesting and telling example of the complexity of ethnic relations during this transitional time period. In 1677 the Apalachee designed and implemented an attack on a village of Chisca, Chacato, and Sabacola who had been raiding the Apalachee villages at night. Because the Apalachee were unfamiliar with the territory, a group of Chacato (who had emigrated to Apalachee and became Christians) guided the raid but the majority refused to participate and exact violence against other Chacato.

13. Consider for example these two quotes: "The introduction of firearms among the Indians of North America drastically altered relationships between rival tribes and resulted in dislocations and forced migrations" (Worcester and Schilz, 1984: 103) versus "this fact [that Spanish mission Indians had firearms] tends to discredit the myth that northern Indians armed with English guns ... were primarily responsible for the population dislocations of the late seventeenth century" (Waselkov, 1989: 121).

14. At the Convento de San Francisco in St. Augustine, a distinctively European context, Hoffman (1993) reported very few items attributed to the unspecified category of "arms": Only 17 of 6,390 items were attributed to this category, and in 1702 at the height of tensions between England and Spain, only 10 of 2,960 items

were assigned to this artifact class. Interestingly, however, the frequency of arms did increase throughout the 17th century, in concert with expectations based on hemispheric politics and the inventories of the St. Augustine garrison presented by Brown (1980). Other excavated segments of the St. Augustine community also produced firearms technology. There was one musket ball in addition to lead clumps at the Joseph de León site (SA-26-1) (Deagan, 1978b), a matchlock musket lock was found at the Trinity Episcopal Church site (SA-34-1)(Deagan, 1978b), and lead shot and a musket lock were found at the María de la Cruz site (Deagan, 1983: 115). At the Fountain of Youth site, Seaberg (1991) reported the presence of material culture associated with arms and weaponry.

Several sites in Timucua province also produced evidence of munitions and firearms. Loucks (1979) summarized the existing information on European weaponry in her dissertation which has since been updated by additional case reports from more recent excavations. Loucks (1993) reported the recovery of lead shot from San Agustín de Urica (Baptizing Spring), noting that two of four pieces were from an aboriginal context. The dearth of shot (4 of 462 European artifacts) suggests its rarity. Loucks (1979) also indicates the presence of gunflints at this mission, which importantly, is thought to predate the 1656 Timucua rebellion (Loucks, 1993). Weisman (1993) also found lead shot and gunflints at the pre-1656 San Martín de Timucua site. Although, as at San Agustín de Urica, the density of lead shot was low (12 of 5,403 total artifacts), over 80 percent of it was found in aboriginal contexts. The ratio of lead shot to projectile points was about 1:20 (12 versus 225), which may suggest something about prevalence and access. Interestingly, one piece of shot was apparently fired at the church (Weisman, 1992: 61), which may reflect San Martín's destruction during the 1656 rebellion. Loucks (1979) also reports the recovery of musket parts from this site and Deagan (1972: 36) reported finding 18 pieces of shot of a caliber consistent with hunting. A broad inventory of material from other sites in Timucua province indicates similar prevalence information. Musket parts, gunflints, and lead shot were recovered at San Juan del Puerto, and a ramrod tip, a gunflint, and lead shot were recovered from the Fox Pond site, thought to be San Francisco de Potano (Loucks, 1993: 307; Symes and Stephens, 1965).

Similar materials were recovered from sites in Apalachee province. Arms recovered from the Spanish village at San Luis de Talimali included gunflints, lead shot, and lead pellets, all found in trash pits (McEwan, 1993b). Despite the number of soldiers in residence at San Luis, the overall density of arms was low, comprising 26 items in total. Shapiro and McEwan (1992: 56–57) reported a partial matchlock firing mechanism, five flints, five pieces of lead shot, and 15 pieces of bird shot within the Apalachee council house (see Hann and McEwan, 1998: 116). Excavations at the fort also produced weaponry, primarily gunflints and lead shot (McEwan and Poe, 1994: 104). The *convento*, a distinctly European structure, produced four gunflints and seven pieces of lead shot (Shapiro and Vernon, 1992). Gunflints were found at San Joseph de Ocuya (Jones, 1973), and Brown (1980: 167) presents a matchlock musket rest from San Damián de Escambé and a partial firing mechanism from

San Juan de Aspalaga. Musket parts, gunflints, and lead shot were recovered from the Scott Miller site, thought to be San Francisco de Oconee (Loucks, 1979) or La Concepción de Ayubale (Jones and Shapiro, 1990). Interestingly, no weaponry was reported from the early 17th-century Apalachee mission San Pedro y San Pablo de Patale (Jones et al., 1991; Marrinan, 1991) or at the O'Connell mission site where the Patale congregation moved after 1650 (Marrinan et al., 2000).

15. For example, the 1677 Apalachee assault against the Chisca included 30 harquebusiers among 190 warriors (Hann, 1988: 148, 186), which Hann has interpreted as a lack of reliance on guns. In 1686, the Delgado expedition into the Creek interior included 20 Apalachee with firearms and another 20 with bows (Boyd, 1937: 10). In 1700 Alonso de Leturiondo noted about the Apalachee, "their arms are the bow and arrow, which they handle with great skill. And today they use firearms as do the Spaniards, and, in Apalache, they maintain their arms as well as do the best trained officers" (Leturiondo in Hann, 1986b: 199). In 1704, the Apalachee produced 93 firearms in defense of the province (Boyd et al., 1951: 52). The Guale were also outfitted with firearms in their defense of St. Catherines Island against the 1680 Westo raid (Swanton, 1922: 90).

16. For example, in reference to the 1680 raid on Guale province, Governor Hita Salazar noted that "on this occasion I have considered it important that the Indians have them [firearms]" (Hita Salazar 1680 in Worth, 1995: 31). This parenthetical clearly implies an element of control. In addition, Governor Rebolledo had previously seized all firearms in Guale fearing an uprising similar to that which occurred in Timucua in 1656 (Worth, 1992: 244–246) and Bushnell (1981: 51) indicates that the treasury of the colony was responsible for tracking the locations of small firearms.

17. Enumerations of the armory of St. Augustine indicate a prevalence of the poorer quality matchlock muskets and arquebuses. In 1653, before conflict with England and slave raiding intensified, the armory of St. Augustine contained 178 matchlocks and six arquebuses, all of which did not work (Chatelain, 1941: 56). As a testament to the slow Spanish response to the English threat, the matchlock continued to dominate the armory in St. Augustine and there were no listed flintlocks until 1683 (Brown, 1980: 115–118). It was not until 1694 that the number of flintlock weapons outnumbered matchlock muskets (Brown, 1980: 118); despite this, Moore's 1702 assault on St. Augustine was met with a force comprising 62 percent matchlock musketeers and arquebusiers with no flintlocks, and the rest pikemen (Arnade, 1959: 39).

18. Peterson (1956: 16–17) provides a nice description of the labor involved in firing a matchlock firearm: "The piece was then loaded, but several actions were still necessary before it could be fired. The match had to be returned to the serpentine and adjusted. The coal on its end had to be blown into activity. If the gunner was forced to wait any length before firing, he had to change the adjustment of the match continually to insure that it would strike the pan and also to prevent it from burning down to the serpentine and going out." Hahn (1995) also provides a concise summary of weaponry technology in the context of southeastern colonial warfare.

As just one example of the deficiency of the matchlocks in combat Brown (1980: 115) offered the following fascinating anecdote from the Robert Searle's nighttime assault on St. Augustine: "Several *arcubuceros* were slain when the pirates, armed with flintlock firearms, detected their glowing match."

19. After the Westo began wreaking havoc in the Georgia interior and along the South Carolina coast, a number of groups began migrating to the Spanish vicinities to seek the protection of the soldiers in residence. The Yamassee and Escamaçu as well as residents of "a variety of others [*sic*] provinces or towns which are either unnamed, or which have names of as yet undetermined origin" (Worth, 1995: 30), began descending on the Spanish mission of the Georgia coast soon after the Westo assaults in the Georgia interior began (ca. 1659). Groups from as far away as central Georgia (Tama) joined the mix. They were welcomed by the Spanish as a source of labor, and by 1675 the number of non-Christian residents of Guale and Mocama (a Timucua district just south of Guale) outnumbered the Catholic population (Worth, 1995: 28). Yamassee were also responsible for repopulating the Mayaca region missions of San Salbador de Mayaca and San Antonio Anacape in 1679 (Hann, 1993b: 122) and made an appearance in Apalachee province at about the same time (Hann, 1988). Apalachee, in particular, experienced significant in-migration of a number of distinct ethnic groups as a direct result of Westo depredations. Individuals identified as Chine, Chacato, Pacara, Amacano, and Oconee were all recorded in settlements within the Apalachee district during the latter half of the seventeenth century (Hann, 1988: 102, 103, 165, 179, 322; 2003: 121–122). Census data suggest 8–10 percent of the population of Apalachee (649 individuals) was comprised of nonlocal ethnic groups. Large numbers of Timucua were also living in Apalachee after the 1656 rebellion (Hann, 1986a: 102, 103) and the entire second half of the 17th century witnessed a veritable melting pot of ethnic groups moving into and out of the Spanish sphere of influence. Interestingly, the Timucuan interior did not seem to have been an epicenter of immigration by allochthonous populations from central Georgia (Hann, 1996a). This may reflect the fact that this part of Florida had little to offer immigrants and was, in fact, rife with fugitivism by local populations. If anything, a movement of Yustaga to missions farther east is most likely responsible for the influx of people along the *camino real* in central Florida (Hann, 1986a, 1988, 1996a).

20. I may be reading Ferguson and Whitehead (1992) too literally here. They do clearly indicate that the tribal zone is not under direct administration by a state entity and for this reason the colony of La Florida would not itself likely constitute an appropriate sphere of interaction. Reviews of their book, however, differ in how the tribal zone is conceptualized. For example, Redmond (2001: 551) defined tribal zones as "regions where such state-nonstate contact occurs," which produces a less restricted meaning. In the preface to the second edition, however, Ferguson and Whitehead (2000) clearly consider the tribal zone as separate from the state's administrative territory and, in fact, invoke Pratt's (1991) concept of the contact zone as a homologous construct.

21. The Chisca were apparently displaced from the Tennessee Valley during the Juan Pardo expeditions and had been raiding Spanish missions since at least 1618 (Hann, 1988: 182–183, 402; Hudson, 1990; Worth, 1998b: 52).

Chapter 6. The Liminal Phase of Ethnogenesis: Practice and the Lived Experience

1. Bourdieu made no explicit connection between habitus and ethnicity and was more interested in the educational system as social construct and the identities of the working class (Jenkins, 1982). More nuanced applications of practice theory in archaeology are likewise not interested in ethnicity but are more interested in historical processual uses of practice theory (e.g., Pauketat, 2000, 2001), practice, and technology, or multi-scalar, multi-attribute applications of practice theory at the community level (e.g., Lightfoot et al., 1998).

2. The Jesuit priest Rogel wrote in 1570, "In order to obtain fruit in the blind and sad souls of these provinces, it is necessary first of all to order the Indians to come together, and live in towns and cultivate the earth" (Gannon, 1965: 33).

3. At the time of Deagan's writing (1978b), Fig Springs was identified as the post-1660 Timucua mission Santa Catalina de Afuerica; it is now thought to be San Martín de Timucua which was destroyed in 1656 (Weisman, 1992), a date corroborated by the majolica inventory (Weisman, 1993: 171). Therefore, the synchronicity of ceramic changes in Utina and Apalachee noted by Milanich (1978) has been adjusted.

4. This refers to the desire to be buried near a saint or holy ground such as that associated with a church cemetery.

5. Hann, whose work provides the most complete analysis of population size data for Apalachee (Hann, 1986b, 1988), provides two intriguing quotes about the evidence for pre-1650 epidemics in Apalachee province. In reference to the mission lists of 1657 Hann (1986a: 392) concludes, "The 1657 visitation record and the other pertinent documents from the era give no indication that Apalachee was experiencing the calamitous demographic dislocation and decline that was manifest in Timucua." Hann (1988: 175) lists three epidemics known to have affected Apalachee in 1655–1657, 1693, and 1703. The last of these was recorded at San Luis itself, whereas the first two dates refer to general impacts on Apalachee province. The 1655–1657 epidemic affected the entire population of La Florida, whereas the 1703 epidemic is said to have killed 170 at San Luis. Despite this, Hann (1988: 175) comments, "there is surprisingly little documentary evidence about the epidemiological mechanisms by which Apalachee's and Florida's populations were reduced so drastically."

6. As noted by Hann (1994: 338), Apalachee remained relatively free of *repartimiento* demands until after the 1647 rebellion.

Chapter 7. Bridging Histories: Seminole Ethnogenesis Reconsidered

1. Romans (1961: 62) described the Creeks as "a mixture of the remains of the *Cawittas, Talepoosas, Coosas, Apalachias, Conshacs* or *Coosades, Oakmulgis, Oconis,*

Okchoys, Alibamons, Natchez, Weetumkus, Pakanas, Taënsas, Chacsihoomas, Abékas and some other tribes whose names I do not recollect."

2. In 1685 and 1686, Antonio Matheos and Apalachee allies burned the Apalachicola towns of Coweta, Kasihta, Colone, and Tasquiqui (Hann, 1988).

3. Kelton's work suggests that the Yamassee War was in part the result of the recent spate of serious epidemics that affected the Southeast. In his analysis, the 1696 smallpox epidemic was the first major epidemic in the Southeast that was felt across a broad swath of the region (but excluding Florida). This epidemic was followed by a series of "aftershock epidemics" (see Kelton, 2007: 162) that diminished the available supply of populations that the Georgia groups could raid to sell to the English. The Yamassee War, therefore, was a response to these changing economic circumstances.

4. Chief Tonapi of the Apalachee Seminole (near Tallahassee) clearly differentiated the Apalachee and Alachua communities. Tonapi reported in 1777, "He [Tonapi] has news that although some *Cimarrones*, which for their misdeeds had fled from the *Uchises* [Lower Creeks], had been persuaded by the English of St. Augustine to make war, as they did, against the English of Savannah" (Boyd and Latorre, 1953: 113). That Tonapi identified the Alachua groups as Seminoles, while being referred to as Uchises throughout the narrative, is telling of the identity differences between the Seminole groups during the 18th century.

5. Sturtevant (1971) notes that the proto-Seminole acted independently of the Lower Creeks for the 1765 Treaty of Picolata, but the Lower Creeks did represent the proto-Seminole in the Treaty of Pensacola that same year.

6. McGillivray, in particular, was able to maintain some sense of cohesion among the Georgia and Florida bands through his leadership, but he was ultimately a divisive force in his polarization of the Lower Creeks and Seminoles. Bowles was an influential British trader who instigated animosities between Creek and proto-Seminole towns through the manipulation of trading dynamics with the Spanish and English.

7. The ascription of the Florida communities as distinct from the Georgia Creeks occurred prior to the self-identification of a shared cultural identity by the Seminole themselves. As Weisman (2000: 307) noted, "the Seminoles stopped thinking of themselves as politically bound to the Creek Confederacy some seventy years before their cultural identity became defined in its own unique way." The critical dates are 1765, when the Alachua Seminole eschewed the Treaty of Picolata, and 1835, when the Second Seminole War erupted with the rejection of the Treaty of Moultrie Creek (see Weisman, 2000: 308). As noted in chapters 3 and 4, disparities in the timing of self and other identification is fairly common in the calculus of ethnogenesis (see Horowitz, 1975). In Weisman's analysis, Seminole ethnogenesis coalesced due to opposition to the "Other," which created a sense of shared historical fate and as represented in the active signaling of group cohesion through symbolic expressions of traditionalism and the rejection of European goods and culture (see Weisman, 1989: 106–123; 2000). This reasoning is identical to that which I suggest defined ethnogenetic rationalization during the latter half of the 17th century in La Florida.

8. Porter (1949: 367) notes that Brims's policy of vacillation may have been a shrewd measure to never assure the French, Spanish, or English of his allegiance, which increased the flow of gifts into his possession. Indeed Oglethorpe commented in 1741 (after Brims's death; see Porter, 1949: 371) that the Creeks were poor allies and only responded to gifts (TePaske, 1964: 214). Hahn (2004) attributes the ultimate basis for the formation of the Creek Nation to Brims's Coweta Resolution, which astutely used shifting allegiances to ensure the maintenance of a neutralist policy.

9. Sattler (1996: 46) infers these dates based on Tonapi's statements about his learning the use of firearms, the age at which Creek boys were taught to use firearms, and the date that the statement was recorded (in 1777—see Boyd and Latorre, 1953: 110–114; see also Sattler, 1996, note 15).

10. See Porter (1949) for a discussion of the history of Secoffee and his dealings in Creek internal politics. If Secoffee "founded" the Apalachee Seminole, he certainly did not remain in these communities very long as his name appears throughout the annals in different contexts. He died in the 1720s as a result of alcohol consumption (Hahn, 2004: 136). He was Brims's appointed successor (but not his son—Hahn 2004: 96), something the English vigorously opposed due to his Spanish allegiances. See Cline (1974) and Hahn (2004) for very readable narratives of the machinations of Secoffee and Brims's other kin.

11. An abandoned field or village.

12. Swanton (1922: 179) notes an Oconee presence in several missions during the 17th century. One of these, San Francisco de Oconi, was located in Apalachee province. Although Hann (1988: 191–192) remains unconvinced of the ethnic identity of this mission's inhabitants, he does allow for the fact that the Oconee may have migrated to Apalachee prior to the establishment of missions there. Swanton (1922: 179) also locates the Oconee in a village along the Georgia coast, perhaps near Jekyll Island in Guale province.

13. See Hann (1988: 184) for an example involving an ethnic Chisca.

References Cited

Adachi N, Dodo Y, Ohshima N, Doi N, Yoneda M, Matsumara H. 2003. Morphologic and genetic evidence for the kinship of juvenile skeletal specimens from a 2,000-year-old double burial of the Usu-Moshiri site, Hokkaido, Japan. Anthropological Science 111: 347–363.

Albers PC. 1993. Symbiosis, merger, and war: Contrasting forms of intertribal relationship among historic plains Indians. In: Moore JH, ed. The political economy of North American Indians. Norman: University of Oklahoma Press. 94–132.

———. 1996. Changing patterns of ethnicity in the northeastern Plains, 1780–1870. In: Hill JD, ed. History, power, and identity. Ethnogenesis in the Americas, 1492–1992. Iowa City: Iowa City Press. 90–118.

Albers PC, James WR. 1986. On the dialectics of ethnicity: To be or not to be Santee (Sioux). Journal of Ethnic Studies 14: 1–27.

Alvesalo L, Tigerstedt PMA. 1974. Heritabilities of human tooth dimensions. Hereditas 77: 311–318.

Alvord CW, Bidgood L. 1912. The first explorations of the trans-Allegheny region by the Virginians, 1650–1674. Cleveland: Arthur H. Clark.

Anderson DG. 1994a. The Savannah River chiefdoms: Political change in the late prehistoric Southeast. Tuscaloosa: University of Alabama Press.

———. 1994b. Factional competition and the political evolution of Mississippian chiefdoms in the Southeastern United States. In: Brumfiel E, Fox JW, eds. Factional competition and political development in the New World. New York: Cambridge University Press. 61–76.

Armelagos GJ, Van Gerven DP. 2003. A century of skeletal biology and paleopathology: Contrasts, contradictions, and conflicts. American Anthropologist 105: 53–64.

Arnade CW. 1959. The siege of St. Augustine in 1702. University of Florida Monographs Social Sciences, 3. Gainesville: University of Florida Press.

———. 1960. The failure of Spanish Florida. The Americas 16: 271–281.

Arnaiz-Villena A, Elaiwa N, Silvera C, Rostom A, Moscoso J, Gomez-Casado E, Allende L, Varela P, Martinez-Laso J. 2001. The origin of Palestinians and their genetic relatedness with other Mediterranean populations. Human Immunology 62: 889–900.

Arutiunov S. 1994. Ethnogenesis: Its forms and rules. Anthropology and Archaeology of Eurasia 33: 79–93.

Arya R, Duggirala R, Comuzzie AG, Puppala S, Modem S, Busi BR, Crawford MH. 2002. Heritability of anthropometric phenotypes in caste populations of Visakhapatnam, India. Human Biology 74: 325–344.

Bandelier F. 1905. The journey of Alvar Nuñez Cabeza de Vaca and his companions from Florida to the Pacific, 1528–1536. New York: Allerton.

Banks M. 1996. Ethnicity: Anthropological constructions. London: Routledge.

Barbujani G, Sokal RR. 1990. Zones of sharp genetic change in Europe are also linguistic boundaries. Proceedings of the National Academy of Sciences USA 87: 1816–1819.

Barcia A. 1951. Chronological history of the continent of Florida. Gainesville: University of Florida Press.

Barth F. 1969. Introduction. In: Barth F, ed. Ethnic groups and boundaries. The social organization of cultural difference. Oslo: Universitetsforlaget. 9–38.

Bartram W. 1988. Travels through North and South Carolina, Georgia, east and west Florida, the Cherokee country, the extensive territories of the Muscogulges, or Creek confederacy, and the country of the Chactaws. New York: Penguin Books.

Bateman R, Goddard I, O'Grady R, Funk VA, Mooi R, Kress WJ, Cannell P. 1990. Speaking of forked tongues. The feasibility of reconciling human phylogeny and the history of language. Current Anthropology 31: 1–24.

Batista O, Kolman CJ, Bermingham E. 1995. Mitochondrial DNA diversity in the Kuna Amerinds of Panamá. Human Molecular Genetics 4: 921–929.

Bell A. 2005. White ethnogenesis and gradual capitalism: Perspectives from colonial archaeological sites in the Chesapeake. American Anthropologist 107: 446–460.

Bellesiles MA. 1998. Gun laws in early America: The regulation of firearms ownership, 1607–1794. Law and History Review 16: 567–589.

Bellwood P. 1996. Phylogeny vs reticulation in prehistory. Antiquity 70: 881–890.

Bennett CE. 1968. Settlement of Florida. Gainesville: University of Florida Press.

———. 1975. Three voyages. René Laudonnière. Gainesville: University Presses of Florida.

———. 2001. Laudonnière & Fort Caroline. History and documents. Tuscaloosa: University of Alabama Press.

Bentley GC. 1987. Ethnicity and practice. Comparative Studies in Society and History 29: 24–55.

———. 1991. Response to Yelvington. Comparative Studies in Society and History 33: 169–175.

Bilby K. 1996. Ethnogenesis in the Guianas and Jamaica: Two maroon cases. In: Hill, JD, ed. History, power, and identity. Ethnogenesis in the Americas, 1492–1992. Iowa City: University of Iowa Press. 119–141.

Blakely RL. 1988. The King site: Continuity and contact in sixteenth-century Georgia. Athens: University of Georgia Press.

Blangero J. 1990. Population structure analysis using polygenic traits: Estimation of migration matrices. Human Biology 62: 27–48.

Blitz JH. 1999. Mississippian chiefdoms and the fission-fusion process. American Antiquity 64: 577–592.

Boev P. 1980. The anthropological origins of the Thracians. The Mankind Quarterly 20: 321–330.

Bolton HE. 1917. The mission as a frontier institution in the Spanish-American colonies. American Historical Review 23: 42–61.

———. 1921. The Spanish borderlands: A chronicle of old Florida and the Southwest. New Haven: Yale University Press.

Bolton HE, Ross M. 1925. The debatable land. A sketch of the Anglo-Spanish contest for the Georgia country. Berkeley: University of California Press.

Bostwick JA. 1976. Aboriginal ceramics in pre-18th century colonial St. Augustine, Florida: The de Leon site. Conference on Historic Sites Archeology Papers 11: 140–150.

Bourdieu P. 1977. Outline of a theory of practice. Cambridge: Cambridge University Press.

———. 1984. Distinction: A social critique of the judgment of taste. Cambridge: Harvard University Press.

Bourne EG. 1922. Narratives of the career of Hernando de Soto in the conquest of Florida. 2 volumes. New York: Allerton.

Bowden DEJ, Goose DH. 1969. Inheritance of tooth size in Liverpool families. Journal of Medical Genetics 6: 55–58.

Bowne EE. 2000. The rise and fall of the Westo Indians: An evaluation of the documentary evidence. Early Georgia 28: 56–78.

———. 2005. The Westo Indians. Slave traders of the early colonial South. Tuscaloosa: University of Alabama Press.

———. 2006. "A bold and warlike people": The basis of Westo power. In: Pluckhahn TJ, Ethridge R, eds. Light on the path. The anthropology and history of the southeastern Indians. Tuscaloosa: University of Alabama Press. 123–132.

Boyd MF. 1937. The expedition of Marcos Delgado from Apalachee to the Upper Creek country in 1686. Florida Historical Quarterly 16: 3–48.

———. 1949. Diego Peña's expedition to the Apalachee and Apalachicolo in 1716. Florida Historical Quarterly 28: 1–27.

———. 1952. Documents describing the second and third expeditions of Lieutenant Diego Peña to Apalachee and Apalachicolo in 1717 and 1718. Florida Historical Quarterly 31: 109–139.

Boyd MF, Latorre JN. 1953. Spanish interest in British Florida, and in the progress of the American Revolution. Florida Historical Quarterly 32: 92–130.

Boyd MF, Smith HG, Griffin JW. 1951. Here they once stood: The tragic end of the Apalachee missions. Gainesville: University of Florida Press.

Bromley Yu. 1974. The term *ethnos* and its definition. In: Bromley Yu, ed. Soviet ethnology and anthropology today. The Hague: Mouton. 55–72.

Brown JA, Kerber RA, Winters HD. 1990. Trade and the evolution of exchange relations at the beginning of the Mississippian period. In: Smith BD, ed. The Mississippian emergence. Washington, D.C.: Smithsonian Institution Press. 251–280.

Brown ML. 1980. Firearms in colonial America. The impact on history and technology, 1492–1792. Washington, D.C.: Smithsonian Institution Press.

Buikstra JE, Beck LA. 2006. Bioarchaeology: The contextual analysis of human skeletal remains. Boston: Academic Press.

Buikstra JE, Frankenberg SR, Konigsberg LW. 1990. Skeletal biological distance studies in American physical anthropology. American Journal of Physical Anthropology 82: 1–7.

Buncak J, Piscova M. 2000. Modern national identity of Slovaks and their attitude towards Europe. Sociologia 32: 289–310.

Bushnell AT. 1978. The Menéndez Marquéz cattle barony at La Chua and the determinants of economic expansion in seventeenth-century Florida. Florida Historical Quarterly 56: 407–431.

———. 1979. Patricio de Hinachuba: Defender of the word of God, the Crown of the king and the little children of Ivitachuco. American Indian Culture and Research Journal 3: 1–21.

———. 1981. The king's coffer. Proprietors of the Spanish Florida treasury, 1565–1702. Gainesville: University Press of Florida.

———. 1986. Santa Maria in the written record. Miscellaneous Project Report Series 21. Florida State Museum, Department of Anthropology, Gainesville, Florida.

———. 1990. The sacramental imperative: Catholic ritual and Indian sedentism in the provinces of Florida. In: Thomas DH, ed. Columbian consequences, vol. 2: Archaeological and historical perspectives on the Spanish borderlands east. Washington, D.C.: Smithsonian Institution Press. 475–490.

———. 1994. Situado and sabana: Spain's support system for the presidio and mission provinces of Florida. Anthropological Papers of the American Museum of Natural History, no. 74. Athens: University of Georgia Press.

———. 2006. Ruling "the Republic of Indians" in seventeenth-century Florida. In: Waselkov GA, Wood PH, Hatley MT, eds. Powhatan's mantle. Indians in the colonial Southeast. Lincoln: University of Nebraska Press. 195–213.

Bushnell DI. 1920. Native cemeteries and forms of burial east of the Mississippi. Bureau of American Ethnology Bulletin 71. Washington, D.C.: Government Printing Office.

Caldwell J, McCann C. 1941. Irene Mound site, Chatham County, Georgia. Athens: University of Georgia Press.

Canuto MA, Yaeger J. 2000. The archaeology of communities: A New World perspective. London: Routledge.

Carson, EA. 2006. Maximum likelihood estimation of human craniometric heritabilities. American Journal of Physical Anthropology 131: 169–180.

Cavalli-Sforza LL. 1997. Genes, peoples, and languages. Proceedings of the National Academy of Sciences USA 94: 7719–7724.

Cavalli-Sforza LL, Minch E, Mountain JL. 1992. Coevolution of genes and languages revisited. Proceedings of the National Academy of Sciences USA 89: 5620–5624.

Cavalli-Sforza LL, Piazza A, Menozzi P, Mountain J. 1988. Reconstruction of human evolution: Bringing together genetic, archaeological, and linguistic data. Proceedings of the National Academy of Sciences USA 85: 6002–6006.

Chapman M. 1993a. Social and biological aspects of ethnicity. In: Chapman M, ed. Social and biological aspects of ethnicity. Oxford: Oxford University Press. 1–46.

———. 1993b. Social and biological aspects of ethnicity. Oxford: Oxford University Press.

Chatelain VE. 1941. The defenses of Spanish Florida 1565 to 1763. Washington, D.C.: Carnegie Institution of Washington Publication 511.

Cheves L. 1897. The Shaftesbury papers and other records relating to Carolina and the first settlement on the Ashley River prior to the year 1676. Collections no 5. South Carolina Historical Society, Charleston.

Chropovsky B. 1982. The ethnogenesis of the Slovaks. Historicky Casopsis 30: 19–27.

Cline HF. 1974. Notes on colonial Indians and communities in Florida, 1700–1821. New York: Garland.

Cohen AP. 1969. Custom and politics in urban Africa: A study of Hausa migrants in Yoruba towns. London: Routledge.

———. 1974. Urban ethnicity. London: Tavistock.

———. 1985. The symbolic construction of community. London: Routledge.

Cohen R. 1978. Ethnicity: Problem and focus in anthropology. Annual Review of Anthropology 7: 379–403.

Collard M, Shennan SJ. 2000. Processes of culture change in prehistory: A case study from the European Neolithic. In: Renfrew C, Boyle K, eds. Archaeogenetics: DNA and the population prehistory of Europe. Cambridge: McDonald Institute for Archaeological Research. 89–97.

Collard M, Shennan SJ, Tehrani JJ. 2006. Branching, blending, and the evolution of cultural similarities and differences among human populations. Evolution and Human Behavior 27: 169–184.

Collard M, Tehrani J. 2005. Phylogenesis versus ethnogenesis in Turkmen cultural evolution. In: Mace R, Holden CJ, Shennan SJ, eds. The evolution of cultural diversity: A phylogenetic approach. London: UCL Press. 109–132.

Comaroff JL. 1987. Of totemism and ethnicity: Consciousness, practice and the signs of inequality. Ethnos 52: 301–323.

Conkey MW. 1990. Experimenting with style in archaeology: Some historical and theoretical issues. In: Conkey MW, Hastorf CA, eds. The uses of style in archaeology. Cambridge: Cambridge University Press. 5–17.

Connor JT. 1927. Jean Ribaut: The whole & true discouerye of terra Florida. Deland: Florida State Historical Society.

Connor W. 1978. A nation is a nation, is a state, is an ethnic group is a . . . Ethnic and Racial Studies 1: 377–400.

Cook FC. 1980. Aboriginal mortality on the Georgia coast during the early historic period. South Carolina Antiquities 12: 36–42.

Cordell LS, Yannie VJ. 1991. Ethnicity, ethnogenesis, and the individual: A processual approach toward dialogue. In: Preucel RW, ed. Processual and postprocessual archaeologies. Carbondale: Center for Archaeological Investigations, Southern Illinois University-Carbondale. 96–107.

Cornell S. 1988. Structure, content, and logic in ethnic group formation. Working Paper Series, Center for Research on Politics and Social Organization, Department of Sociology, Harvard University.

Corruccini RS, Potter RHY. 1980. Genetic analysis of occlusal variation in twins. American Journal of Orthodontics 78: 140–154.

Corruccini RS, Shimada I. 2002. Dental relatedness corresponding to mortuary patterning at Huaca Loro, Peru. American Journal of Physical Anthropology 117: 113–121.

Corruccini RS, Shimada I, Shinoda K. 2002. Dental and mtDNA relatedness among thousand-year-old remains from Huaca Loro, Peru. Dental Anthropology 16: 9–14.

Covington JW. 1967. Some observations concerning the Florida-Carolina Indian slave trade. Florida Anthropologist 20: 10–18.

———. 1968. Stuart's Town, the Yamasee Indians and Spanish Florida. Florida Anthropologist 21: 8–13.

———. 1972. Apalachee Indians, 1704–1763. Florida Historical Quarterly 50: 366–384.

———. 1993. The Seminoles of Florida. Gainesville: University Press of Florida.

Craig AK, Peebles C. 1974. Ethnoecologic change among the Seminoles, 1740–1840. Geoscience and Man 5: 83–96.

Crane VW. 1918. A historical note on the Westo Indians. American Anthropologist 20: 331–337.

———. 1919. Westo and Chisca. American Anthropologist 21: 463–465.

———. 1956. The southern frontier, 1670–1732. Ann Arbor: University of Michigan Press.

Davis DD. 2001. A case of identity: Ethnogenesis of the New Houma Indians. Ethnohistory 48: 473–494.

Davis TF. 1935. Ponce de Leon's first voyage and discovery of Florida. Florida Historical Quarterly 14: 5–70.

———. 1991. History of Juan Ponce de Leon's voyages to Florida. In: Milanich JT, ed. Spanish borderlands sourcebook. 12. Earliest Hispanic/Native American interactions in the American Southeast. New York: Garland. 3–68.

Deagan KA. 1972. Fig Springs: The mid-seventeenth century in north-central Florida. Historical Archaeology 6: 23–46.

———. 1973. *Mestizaje* in colonial St. Augustine. Ethnohistory 20: 53–65.

———. 1978a. Cultures in transition: Fusion and assimilation among the eastern Timucua. In: Milanich J, Proctor S, eds. Tacachale: Essays on the Indians of Florida and southeastern Georgia during the historic period. Gainesville: University Presses of Florida. 89–119.

———. 1978b. The material assemblage of 16th century Spanish Florida. Historical Archaeology 12: 25–50.

———. 1983. Spanish St. Augustine: The archaeology of a colonial Creole community. New York: Academic Press.

———. 1985. Spanish-Indian interaction in sixteenth century Florida and Hispaniola. In: Fitzhugh WW, ed. Cultures in contact. Washington, D.C.: Smithsonian Institution Press. 281–318.

———. 1990a. Accommodation and resistance: The process and impact of Spanish colonization in the Southeast. In: Thomas DH, ed. Columbian consequences, vol. 2: Archaeological and historical perspectives on the Spanish borderlands east. Washington, D.C.: Smithsonian Institution Press. 297–314.

———. 1990b. Sixteenth-century Spanish-American colonization in the southeastern United States and the Caribbean. In: Thomas DH, ed. Columbian consequences, vol. 2: Archaeological and historical perspectives on the Spanish borderlands east. Washington, D.C.: Smithsonian Institution Press. 225–250.

———. 1993. St. Augustine and the mission frontier. In: McEwan BG, ed. The Spanish missions of La Florida. Gainesville: University Press of Florida. 87–110.

———. 1996. Colonial transformation: Euro-American cultural genesis in the early Spanish-American colonies. Journal of Anthropological Research 52: 135–160.

———. 1997. Cross-disciplinary themes in the recovery of the colonial middle period. Historical Archaeology 31: 4–9.

———. 1998. Transculturation and Spanish American ethnogenesis: The archaeological legacy of the quincentenary. In: Cusick JG, ed. Studies in culture contact: Interaction, culture change, and archaeology. Carbondale: Center for Archaeological Investigations, Southern Illinois University-Carbondale. 23–43.

———. 2003. Colonial origins and colonial transformations in Spanish America. Historical Archaeology 37: 3–13.

Dempsey PJ, Townsend GC. 2001. Genetic and environmental contributions to variation in human tooth size. Heredity 86: 685–693.

Dempsey PJ, Townsend GC, Martin NG, Neale MC. 1995. Genetic covariance structure of incisor crown size in twins. Journal of Dental Research 74: 1389–1398.

DePratter CB. 1991. Late prehistoric and early historic chiefdoms in the southeastern United States. New York: Garland.

Derenko MV, Shields GF. 1998. Variation of mitochondrial DNA in three groups of indigenous populations of northern Asia. Genetika 34: 676–681.

De Vos G. 1975. Ethnic pluralism: Conflict and accommodation. In: De Vos G, Romanucci-Ross L, eds. Ethnic identity: Cultural continuities and change. Palo Alto, Calif.: Mayfield. 5–41.

Dobyns HF. 1983. Their number become thinned: Native American population dynamics in eastern North America. Knoxville: University of Tennessee Press.

———. 1991. The invasion of Florida: Disease and the Indians of Florida. In:

Henderson AL, Mormino GR, eds. Spanish pathways in Florida: 1492–1992. Sarasota: Pineapple Press. 58–77.

Douglass WA. 1969. Death in Murelaga: Funerary ritual in a Spanish Basque village. Seattle: University of Washington Press.

Dragadze T. 1980. The place of "ethnos" theory in Soviet anthropology. In: Gellner E, ed. Soviet and western anthropology. New York: Columbia University Press. 161–170.

Dye DH. 1990. Warfare in the sixteenth-century Southeast: The de Soto expedition in the interior. In: Thomas DH, ed. Columbian consequences, vol. 2: Archaeological and historical perspectives on the Spanish borderlands east. Washington, D.C.: Smithsonian Institution Press. 211–222.

———. 2002. Warfare in the protohistoric Southeast, 1500–1700. In: Wesson CB, Rees MA, eds. Between contacts and colonies. Archaeological perspectives on the protohistoric Southeast. Tuscaloosa: University of Alabama Press. 126–141.

Edgar JHJ. 2004. Dentitions, distance, and difficulty: A comparison of two statistical techniques for dental morphological data. Dental Anthropology Journal 17: 55–62.

Edgar W. 1998. South Carolina: A history. Columbia: University of South Carolina Press.

Ehrmann WW. 1940. The Timucua Indians of sixteenth century Florida. Florida Historical Quarterly 18: 168–191.

Emberling G. 1997. Ethnicity in complex societies: Archaeological perspectives. Journal of Archaeological Research 5: 295–344.

Epstein AL. 1978. Ethos and identity: Three studies in ethnicity. London: Tavistock.

Eriksen TH. 2002. Ethnicity and nationalism: Anthropological perspectives, 2nd edition. London: Pluto Press.

Espiritu YL. 1992. Asian American panethnicity: Bridging institutions and identities. Philadelphia: Temple University Press.

Ethridge R. 1984. Flintlocks and slave-catchers: Economic transformations of the Indians of Georgia. Early Georgia 10: 13–26.

———. 2006. Creating the shatter zone: Indian slave traders and the collapse of the southeastern chiefdoms. In: Pluckhahn TJ, Ethridge R, eds. Light on the path. The anthropology and history of the southeastern Indians. Tuscaloosa: University of Alabama Press. 207–218.

Ezzo JA, Larsen CS, Burton JH. 1995. Elemental signatures of human diets from the Georgia Bight. American Journal of Physical Anthropology 98: 471–481.

Fairbanks CH. 1974. Ethnohistorical report of the Florida Indians. New York: Garland.

———. 1978. The ethno-archaeology of the Florida Seminole. In: Milanich J, Proctor S, eds. Tacachale: Essays on the Indians of Florida and southeastern Georgia during the historic period. Gainesville: University Presses of Florida. 163–193.

Ferguson RB. 1984. Warfare, culture, and environment. Orlando: Academic Press.

Ferguson RB, Whitehead NL. 1992. War in the tribal zone. Expanding states and indigenous warfare. Santa Fe: School of American Research.

———. 2000. War in the tribal zone. Expanding states and indigenous warfare, second edition. Santa Fe: School of American Research.
Filipsson R, Goldson L. 1963. Correlation between tooth width, width of the head, length of the head, and stature. Acta Odontologica Scandinavica 21: 359–365.
Fitch T. 1916. Captain Fitch's journal to the Creeks, 1725. In Mereness ND, ed. Travels in the American colonies. New York: Macmillan. 175–212.
Fresia AE, Ruff CB. 1987. Temporal decline in bilateral asymmetry of the upper limb on the Georgia coast. American Journal of Physical Anthropology 72: 199.
Fresia AE, Ruff CB, Larsen CS. 1990. Temporal decline in bilateral asymmetry of the upper limb on the Georgia coast. In: Larsen CS, ed. The archaeology of mission Santa Catalina de Guale: 2. Biocultural interpretations of a population in transition. New York: Anthropological Papers of the American Museum of Natural History, no. 68. 121–132.
Fried M. 1975. The notion of tribe. Menlo Park, Calif.: Cummings.
Gallay A. 2002. The Indian slave trade: The rise of the English empire in the American South, 1670–1717. New Haven, Conn.: Yale University Press.
Gannon MV. 1965. The cross in the sand: The early Catholic church in Florida, 1513–1870. Gainesville: University of Florida Press.
———. 1992. The new alliance of history and archaeology in the eastern Spanish borderlands. William and Mary Quarterly 49: 321–334.
Gardner WM. 1966. The Waddells Mill Pond site. Florida Anthropologist 19: 43–60.
Garn SM, Lewis AB, Kerewsky RS. 1965a. Genetic, nutritional, and maturational correlates of dental development. Journal of Dental Research 44: 228–242.
———. 1965b. Size interrelationships of the mesial and distal teeth. Journal of Dental Research 44: 350–354.
———. 1967a. Sex difference in tooth shape. Journal of Dental Research 46: 1470.
———. 1968. Relationship between buccolingual and mesiodistal tooth diameters. Journal of Dental Research 47: 495.
Garn SM, Lewis AB, Swindler DR, Kerewsky RS. 1967b. Genetic control of sexual dimorphism in tooth size. Journal of Dental Research 5 (supplement): 963–972.
Gatschet AS. 1969. A migration legend of the Creek Indians, with a linguistic, historic and ethnographic introduction. Volume 1. New York: AMS Press.
Geertz C. 1963. The integrative revolution: Primordial sentiments and civil politics in the new states. The quest for modernity in Asia and Africa. In: Geertz C, ed. Old societies and new states. New York: Free Press. 105–157.
Geiger M. 1937. The Franciscan conquest of Florida (1573–1618). PhD dissertation, Catholic University of America.
Gibson JL. 1974. Aboriginal warfare in the protohistoric Southeast: An alternative perspective. American Antiquity 39: 130–133.
Given BJ. 1994. A most pernicious thing. Gun trading and Native American warfare in the early contact period. Ottawa: Carleton University Press.
Glazer N, Moynihan DP. 1975. Introduction. In: Glazer N, Moynihan DP, eds. Ethnicity: Theory and experience. Cambridge: Harvard University Press. 1–28.

Gleach FW. 1997. Powhatan's world and colonial Virginia. A conflict of cultures. Lincoln: University of Nebraska Press.

Goggin JM. 1953. An introductory outline of Timucuan archaeology. Southeastern Archaeological Conference Newsletter 3: 4–17.

Goggin JM, Sturtevant WC. 1964. The Calusa: A stratified, nonagricultural society (with notes on sibling marriage). In: Goodenough WH, ed. Explorations in cultural anthropology: Essays in honor of George Peter Murdock. New York: McGraw Hill. 179–219.

Gonzalez NL. 1988. Sojourners of the Caribbean: Ethnogenesis and ethnohistory of the Garifuna. Urbana: University of Illinois Press.

González-José R, Dahinten SL, Luis MA, Hernández M, Pucciarelli HM. 2001. Craniometric variation and the settlement of the Americas: Testing hypotheses by means of R-matrix and matrix correlation analyses. American Journal of Physical Anthropology 116: 154–165.

Goose DH. 1963. Dental measurement: An assessment of its value in anthropological studies. In: Brothwell DR, ed. Dental anthropology. New York: Pergamon Press. 125–148.

———. 1971. The inheritance of tooth size in British families. In: Dahlberg AA, ed. Dental morphology and evolution. Chicago: University of Chicago Press. 144–149.

Gould SJ. 1996. The mismeasure of man. New York: WW Norton.

Gowland R, Knüsel C. 2006. Social archaeology of funerary remains. Oxford: Oxbow.

Granberry J. 1993. A grammar and dictionary of the Timucua language. Tuscaloosa: University of Alabama Press.

Green W, DePratter CB, Southerlin B. 2002. The Yamassee in South Carolina: Native American adaptation and interaction along the Carolina frontier. In: Joseph JW, Zierden M, eds. Another's country: Archaeological and historical perspectives on cultural interactions in the southern colonies. Tuscaloosa: University of Alabama Press. 13–29.

Griffin MC, Larsen CS. 1989. Patterns in osteoarthritis: A case study from the prehistoric and historic southeastern U.S. Atlantic coast. American Journal of Physical Anthropology 78: 232.

Guglielmino CR, Viganotti C, Hewlett B, Cavalli-Sforza LL. 1995. Cultural variation in Africa: Role of mechanisms of transmission and adaptation. Proceedings of the National Academy of Sciences USA 92: 7585–7589.

Haas J. 1990. The anthropology of war. Cambridge: Cambridge University Press.

Hahn SC. 1995. A miniature arms race: The role of the flintlock in initiating Indian dependency in the colonial southeastern United States, 1656–1730. MA thesis, University of Georgia.

———. 2000. The invention of the Creek Nation. PhD dissertation, Emory University.

———. 2002. The mother of necessity: Carolina, the Creek Indians, and the making of a new order in the American Southeast, 1670–1763. In: Ethridge R, Hudson

C, eds. The transformation of the southeastern Indians, 1540–1760. Jackson: University of Mississippi Press. 79–114.

———. 2004. The invention of the Creek nation, 1670–1763. Lincoln: University of Nebraska Press.

———. 2006. The Cussita migration legend. History, ideology, and the politics of mythmaking. In: Pluckhahn TJ, Ethridge R, eds. Light on the path. The anthropology and history of the southeastern Indians. Tuscaloosa: University of Alabama Press. 57–93.

Hakluyt R. 1810. Collection of the early voyages, travels, and discoveries, of the English nation. London: printed for RH Evans, J Mackinlay, R Priestley.

Haley BD, Wilcoxon LR. 2005. How Spaniards became Chumash and other tales of ethnogenesis. American Anthropologist 107: 432–445.

Hallgrímsson B, Ó Donnabháin B, Walters GB, Cooper DML, Guðbjartsson D, Stefánsson K. 2004. Composition of the founding population of Iceland: Biological distance and morphological variation in early historic Atlantic Europe. American Journal of Physical Anthropology 124: 257–274.

Hally DJ. 1979. Archaeological investigations of the Little Egypt site (9MU102), Murray County, Ga., 1969 season. University of Georgia, Laboratory of Archaeology Series, Report no. 18. Athens.

———. 1980. Archaeological investigation of the Little Egypt site (9MU102), Murray County, Georgia, 1970–72 seasons. Unpublished report submitted to the National Parks Service.

———. 1996. Platform mound construction and the instability of Mississippian chiefdoms. In: Scarry J, ed. Political structure and change in the prehistoric southeastern United States. Gainesville: University Press of Florida. 2–127.

———. 2006. The nature of Mississippian regional systems. In: Pluckhahn TJ, Ethridge R, eds. Light on the path. The anthropology and history of the southeastern Indians. Tuscaloosa: University of Alabama Press. 26–42.

Handelman D. 1977. The organization of ethnicity. Ethnic Groups 1: 187–200.

Hann JH. 1986a. Demographic patterns and changes in mid-seventeenth century Timucua and Apalachee. Florida Historical Quarterly 64: 371–392.

———. 1986b. Spanish translations. Florida Archaeology, no. 2. Tallahassee: Florida Bureau of Archaeological Research.

———. 1987. Twilight of the Mocamo and Guale aborigines as portrayed in the 1695 Spanish visitation. Florida Historical Quarterly 66: 1–24.

———. 1988. Apalachee: The land between the rivers. Gainesville: University Presses of Florida.

———. 1991. Missions to the Calusa. Gainesville: University of Florida Press.

———. 1993a. Visitations and revolts in Florida, 1656–1695. Florida Archaeology, no. 7. Tallahassee: Florida Bureau of Archaeological Research.

———. 1993b. The Mayaca and Jororo and missions to them. In: McEwan BG, ed. The Spanish missions of La Florida. Gainesville: University Press of Florida. 111–140.

———. 1994. The Apalachee of the historic era. In: Hudson C, Tesser CC, eds. The forgotten centuries. Indians and Europeans in the American South, 1521–1704. Athens: University of Georgia Press. 327–354.

———. 1995. The Indian village on Apalachee Bay's Rio Chachave on the Solana Map of 1683. Florida Anthropologist 48: 61–66.

———. 1996a. A history of the Timucua Indians and missions. Gainesville: University Press of Florida.

———. 1996b. Late seventeenth-century forebears of the Lower Creeks and Seminoles. Southeastern Archaeology 15: 66–80.

———. 2003. Indians of central and south Florida, 1513–1763. Gainesville: University Press of Florida.

Hann JH, McEwan BG. 1998. The Apalachee Indians and mission San Luis. Gainesville: University Press of Florida.

Hardesty DL. 1999. Historical archaeology in the next millennium: A forum. Historical Archaeology 33: 51–58.

Hardin, KW. 1986. The Santa Maria mission project. Florida Anthropologist 39: 75–83.

Harpending HC, Jenkins T. 1973. Genetic distance among southern African populations. In: Crawford MH, Workman PL, eds. Method and theory in anthropological genetics. Albuquerque: University of New Mexico Press. 177–199.

Harpending HC, Ward R. 1982. Chemical systematics and human populations. In: Nitecki M, eds. Biochemical aspects of evolutionary biology. Chicago: University of Chicago Press. 213–256.

Harris EF, Smith RJ. 1980. A study of occlusion and arch widths in families. American Journal of Orthodontics 78: 155–163.

———. 1982. Occlusion and arch size in families: A principal components analysis. Angle Orthodontist 52: 135–143.

Harzer W. 1995. Size relationship and interactions between crown diameters in twins and family members. In: Radlanski RJ, Renz H, eds. Proceedings of the 10th international symposium on dental morphology. Berlin. 124–135.

Hauser G, De Stefano GF. 1989. Epigenetic variants of the human skull. Stuttgart: Schweizerbart.

Hawks J. 2008. From genes to numbers: Effective population sizes in human evolution. In: Bocquet-Appel J-P, ed. Recent advances in palaeodemography: Data, techniques, patterns. Dordrecht: Springer Verlag.

Hegmon M. 1992. Archaeological research on style. Annual Review of Anthropology 21: 517–536.

———. 1998. Technology, style, and social practices: Archaeological approaches. In: Stark MT, ed. The archaeology of social boundaries. Washington, D.C.: Smithsonian Institution Press. 264–279.

Hemphill BE, Mallory JP. 2004. Horse-mounted invaders from the Russo-Kazakh steppe or agricultural colonists from western central Asia? A craniometric investigation of the Bronze Age settlement of Xinjiang. American Journal of Physical Anthropology 124: 199–222.

Henderson AM, Corruccini RS. 1976. Relationship between tooth size and body size in American Blacks. Journal of Dental Research 55: 94–96.

Henige D. 1986. Primary source by primary source? On the role of epidemics in New World depopulation. Ethnohistory 33: 293–312.

———. 1997. "So unbelievable it has to be true": Inca Garcilaso in two worlds. In: Galloway P, ed. The Hernando de Soto expedition: History, historiography, and "discovery" in the Southeast. Lincoln: University of Nebraska Press. 155–180.

———. 1998. Numbers from nowhere: The American Indian contact population debate. Norman: University of Oklahoma Press.

Hickerson NP. 1996. Ethnogenesis in the south Plains: Jumano to Kiowa? In: Hill JD, ed. History, power, and identity. Ethnogenesis in the Americas, 1492–1992. Iowa City: University of Iowa Press. 70–89.

Hicks GL. 1977. Introduction: Problems in the study of ethnicity. In: Hicks GL, Leis PE, eds. Ethnic encounters. North Scituate, Maine: Duxbury Press. 1–20.

Hill C, Soares P, Mormina M, Macauley V, Meehan W, Blackburn J, Clarke D, Raja JM, Ismail P, Bulbeck D, Oppenheimer S, Richards M. 2006. Phylogeography and ethnogenesis of aboriginal Southeast Asians. Molecular Biology and Evolution 23: 2480–2491.

Hill JD. 1996a. History, power and identity. Ethnogenesis in the Americas, 1492–1992. Iowa City: University of Iowa Press.

———. 1996b. Introduction: Ethnogenesis in the Americas, 1492–1992. In: Hill JD, ed. History, power and identity. Ethnogenesis in the Americas, 1492–1992. Iowa City: University of Iowa Press. 1–19.

Hillson S. 1996. Dental anthropology. Cambridge: Cambridge University Press.

Hlusko LJ, Maas M-L, Mahaney MC. 2004. Statistical genetics of molar cusp patterning in pedigreed baboons: Implications for primate dental development and evolution. Journal of Experimental Zoology 302B: 268–283.

Hlusko LJ, Mahaney MC. 2007. Of mice and monkeys: Quantitative genetic analyses of size variation along the dental arcade. In: Bailey SE, Hublin J-J, eds. Dental perspectives on human evolution: State of the art research in dental paleoanthropology. Dordrecht: Springer. 237–245.

Hodder I. 1979. Economic and social stress and material culture patterning. American Antiquity 44: 446–454.

———. 1982. Symbols in action. Cambridge: Cambridge University Press.

Hoffman K. 1993. The archaeology of the Convento de San Francisco. In: McEwan BG, ed. The Spanish missions of La Florida. Gainesville: University Press of Florida. 62–86.

———. 1994. The development of a cultural identity in colonial America: The Spanish-American experience in La Florida. PhD dissertation, University of Florida.

———. 1997. Cultural development in *La Florida*. Historical Archaeology 31: 24–35.

Hoppenbrouwers P. 2006. Such stuff as peoples are made on: Ethnogenesis and the construction of nationhood in medieval Europe. Medieval History Journal 9: 195–242.

Horowitz DL. 1975. Ethnic identity. In: Glazer N, Moynihan DP, eds. Ethnicity: Theory and experience. Cambridge: Harvard University Press. 111–140.

Hoshower LM, Milanich JT. 1993. Excavations in the Fig Springs mission burial area. In: McEwan BG, ed. The Spanish missions of La Florida. Gainesville: University Press of Florida. 217–243.

Houghton P. 1996. People of the great ocean. Aspects of human biology of the early Pacific. Cambridge: Cambridge University Press.

Howells WW. 1989. Skull shapes and the map. Craniometric analyses in the dispersion of modern Homo. Papers of the Peabody Museum of Archaeology and Ethnology Harvard University, volume 79.

———. 1995. Who's who in skulls. Ethnic Identification of crania from measurements. Papers of the Peabody Museum of Archaeology and Ethnology Harvard University, volume 82.

Hu JR, Nakasima A, Takahama Y. 1991. Heritability of dental arch dimensions in humans. Journal of Craniofacial Genetics and Developmental Biology 11: 165–169.

Hudson C. 1990. The Juan Pardo expeditions: Exploration of the Carolinas and Tennessee, 1566–1568. Washington, D.C.: Smithsonian Institution Press.

Hudson C, Smith M, DePratter C. 1984. The Hernando de Soto expedition: From Apalachee to Chiaha. Southeastern Archaeology 3: 65–77.

Hudson CM. 1988. A Spanish-Coosa alliance in sixteenth-century Georgia. The Georgia Historical Quarterly 72: 599–626.

Hudson MJ. 1999. Ruins of identity: Ethnogenesis in the Japanese Islands. Honolulu: University of Hawai'i Press.

Hulse FS. 1941. The people who lived at Irene. In: Caldwell J, McCann C, eds. Irene Mound site, Chatham County, Georgia. Athens: University of Georgia Press. 57–68.

Hunley K, Long JC. 2005. Gene flow across linguistic boundaries in native North American populations. Proceedings of the National Academy of Sciences USA 102: 1312–1317.

Hunley KL, Cabana GS, Merriwether DA, Long JC. 2007. A formal test of linguistic and genetic coevolution in native Central and South America. American Journal of Physical Anthropology 132: 622–631.

Hurles ME, Sykes BC, Jobling MA, Forster P. 2005. The dual origin of the Malagasy in island Southeast Asia and East Africa: Evidence from maternal and paternal lineages. American Journal of Human Genetics 76: 894–901.

Huscher HA. 1972. Archaeological investigations in the West Point dam area: A preliminary report. Ms. on file, Department of Anthropology, University of Georgia, Athens.

Hussey RD. 1932. Text of the Laws of Burgos (1512–1513) concerning the treatment of the Indians. Hispanic American Historical Review 12: 301–326.

Hutchinson DL. 1986. Enamel hypoplasia: Biological stress and lifeway on St. Catherines Island, Georgia. MA thesis, Northern Illinois University.

———. 1993. Analysis of skeletal remains from the Tierra Verde site, Pinellas County, west-central Florida. Florida Anthropologist 46: 263–276.

———. 1996. Brief encounters: Tatham Mound and the evidence for Spanish and Native American confrontation. International Journal of Osteoarchaeology 6: 51–65.

———. 2006. Tatham Mound and the bioarchaeology of European contact. Gainesville: University Press of Florida.

Hutchinson DL, Larsen CS. 1988. Determination of stress episode duration from linear enamel hypoplasias: A case study from St. Catherines Island, Georgia. Human Biology 60: 93–110.

———. 1990. Stress and lifeway change: The evidence from enamel hypoplasias. In: Larsen CS, ed. The archaeology of mission Santa Catalina de Guale: 2. Biocultural interpretations of a population in transition. New York: Anthropological Papers of the American Museum of Natural History, no. 68. 50–65.

———. 2001. Enamel hypoplasia and stress in La Florida. In: Larsen CS, ed. Bioarchaeology of Spanish Florida: The impact of colonialism. Gainesville: University Press of Florida. 181–206.

Hutchinson DL, Larsen CS, Norr L, Schoeninger MJ. 2000. Agricultural melodies and alternative harmonies in Florida and Georgia. In: Lambert PM, ed. Bioarchaeological studies of life in the age of agriculture: A view from the Southeast. Tuscaloosa: University of Alabama Press. 96–115.

Hutchinson DL, Larsen CS, Schoeninger MJ, Norr L. 1998. Regional variation in the pattern of maize adoption and use in Florida and Georgia. American Antiquity 63: 397–416.

Hutchinson DL, Mitchem JM. 1996. The Weeki Wachee mound, an early contact period mortuary locality in Hernando County, west-central Florida. Southeastern Archaeology 15: 47–65.

Ioviță RP, Schurr TG. 2004. Reconstructing the origins and migrations of diasporic populations: The case of the European gypsies. American Anthropologist 106: 267–281.

Isaacs HR. 1974. Basic group identity: The idols of the tribe. Ethnicity 1: 15–41.

Isajiw WW. 1974. Definitions of ethnicity. Ethnicity 1: 111–124.

———. 1992. Definition and dimensions of ethnicity: A theoretical framework. Paper presented at Joint Canada-United States Conference on the Measurement of Ethnicity, Ottawa, Canada, April 2, 1992.

Isbell WH. 2000. What we should be studying: The "imagined community" and the "natural community" In: Canuto MA, Yaeger J, eds. The archaeology of communities: A New World perspective. London: Routledge. 243–266.

Jenkins R. 1982. Pierre Bourdieu and the reproduction of determinism. Sociology 16: 270–281.

———. 1986. Social anthropological models of inter-ethnic relations. In: Rex J, Mason D, eds. Theories of race and ethnic relations. Cambridge: Cambridge University Press. 170–186.

———. 1992. Pierre Bourdieu. London: Routledge.

———. 1997. Rethinking ethnicity: Arguments and explorations. London: Sage.

———. 2004. Social identity. London: Routledge.

Jernvall J, Jung H-S. 2000. Genotype, phenotype, and developmental biology of molar tooth characters. Yearbook of Physical Anthropology 43: 171–190.

Johnson KW. 1993. Mission Santa Fé de Toloca. In: McEwan BG, ed. The Spanish missions of La Florida. Gainesville: University Press of Florida. 141–164.

Johnson KW, Nelson BC. 1990. The Utina: Seriations and chronology. Florida Anthropologist 43: 48–62.

Jones BC. 1972. Spanish mission sites located and test excavated. Archives and History News 3: 1–2.

———. 1973. A semi-subterranean structure at mission San Joseph de Ocuya, Jefferson County, Florida. Tallahassee: Bureau of Historic Sites and Properties Bulletin no. 3. 1–50.

———. 1982. Southern cult manifestations at the Lake Jackson site, Leon County, Florida: Salvage excavation of mound 3. Mid-continental Journal of Archaeology 7: 3–44.

———. 1991. Evaluation of human burial site at Holy Spirit Catholic church site in Duval County, Florida. Unpublished ms. on file with author.

Jones BC, Shapiro GN. 1990. Nine mission sites in Apalachee. In: Thomas DH, ed. Columbian consequences, vol. 2: Archaeological and historical perspectives on the Spanish borderlands east. Washington, D.C.: Smithsonian Institution Press. 491–510.

Jones BC, Storey R, Widmer RJ. 1991. The Patale cemetery: Evidence concerning the Apalachee mission mortuary complex. In: Jones BC, Hann J, Scarry JF, eds. San Pedro y San Pablo de Patale: A seventeenth-century Spanish mission in Leon County, Florida. Florida Archaeology, no. 5. Tallahassee: Florida Bureau of Archaeological Research. 109–125.

Jones GD. 1978. The ethnohistory of the Guale coast through 1684. In: Thomas DH, Jones GD, Durham RS, Larsen CS, eds. The anthropology of St. Catherines Island. 1. Natural and cultural history. Anthropological Papers 55, pt. 2. New York: American Museum of Natural History. 178–210.

———. 1980. Guale Indians of the southeastern United States coast. In: Howard JD, DePratter CB, Frey RW, eds. Excursions in southeastern geology: The archaeology-geology of the Georgia coast. Atlanta: Geological Society of America. 215–224.

Jones S. 2002. The archaeology of ethnicity. London: Routledge.

———. 2007. Discourses of identity in the interpretation of the past. In: Insoll T, ed. The archaeology of identities: A reader. London: Routledge. 44–58.

Juricek JT. 1964. The Westo Indians. Ethnohistory 11: 134–173.

Kangas AT, Evans AR, Thesleff I, Jernvall J. 2004. Nonindependence of mammalian dental characters. Nature 432: 211–214.

Kaufmann E. 1999. American exceptionalism reconsidered: Anglo-Saxon ethnogenesis in the "universal" nation, 1776–1850. Journal of American Studies 33: 437–457.

Keller EJ. 1995. The ethnogenesis of the Oromo Nation and its implications for politics in Ethiopia. Journal of Modern African Studies 33: 621–634.

Kelly AR. 1970. Explorations at Bell Field Mound and village. Unpublished report submitted to the National Park Service.

———. 1972. The 1970–71 field seasons at Bell Field Mounds, Carters Dam. Unpublished report submitted to the National Park Service.

Kelly AR, Schnell FT, Smith DF, Schlosser AL. 1965. Explorations in Sixtoe Field, Carter's Dam, Murray County, Georgia. Seasons 1962, 1963, 1964. Unpublished report submitted to the National Park Service.

Kelton P. 2002. The great southeastern Smallpox epidemic, 1696–1700: The region's first major epidemic? In: Ethridge R, Hudson C, eds. The transformation of the southeastern Indians, 1540–1760. Jackson: University Press of Mississippi. 21–37.

———. 2007. Epidemics and enslavement: Biological catastrophe in the Native Southeast 1492–1715. Lincoln: University of Nebraska Press.

Keyes CF. 1981. Ethnic change. Seattle: University of Washington Press.

Kieser JA. 1990. Human adult odontometrics: The study of variation in adult tooth size. Cambridge: Cambridge University Press.

King J. 1984. Ceramic variability in 17th century St. Augustine, Florida. Historical Archaeology 18: 76–82.

Knight VJ, Jr. 1986. The institutional organization of Mississippian religion. American Antiquity 51: 675–687.

———.1994. The formation of the Creeks. In: Hudson C, Tesser CC, eds. The forgotten centuries: Indians and Europeans in the American South, 1521–1704. Athens: University of Georgia Press. 373–392.

Knudson KJ, Stojanowski CM. 2009. Bioarchaeology and identity in the Americas. Gainesville: University Press of Florida.

Kolakowski D, Bailit HL. 1981. A differential environmental effect on human anterior tooth size. American Journal of Physical Anthropology 54: 377–381.

Kolman CJ, Bermingham E. 1997. Mitochondrial and nuclear DNA diversity in the Chocó and Chibcha Amerinds of Panamá. Genetics 147: 1289–1302.

Kolman CJ, Bermingham E, Cooke R, Ward RH, Arias TD, Guionneau-Sinclair F. 1995. Reduced mtDNA diversity in the Ngöbé Amerinds of Panamá. Genetics 140: 275–283.

Konigsberg LW. 2000. Quantitative variation and genetics. In: Stinson S, Bogin B, Huss-Ashmore R, O'Rourke D, eds. Human biology: An evolutionary and biocultural perspective. New York: Wiley-Liss. 135–162.

Konigsberg LW, Buikstra JE. 2006. Population structure analysis from prehistoric skeletal material. American Journal of Physical Anthropology (Supp) 42: 115.

Konigsberg LW, Ousley SD. 1995. Multivariate quantitative genetics of anthropometric traits from the Boas data. Human Biology 67: 481–498.

Kopytoff B. 1976. The development of Jamaican maroon ethnicity. Caribbean Quarterly 22: 33–50.

Kowalewski SA. 1995. Large-scale ecology in aboriginal eastern North America. In: Nassaney MS, Sassaman KE, eds. Native American interactions: Multiscalar analyses and interpretations in the Eastern Woodlands. Knoxville: University of Tennessee Press. 147–173.

Kozintsev A. 1992. Ethnic epigenetics: A new approach. Homo 43: 213–244.
Kupperman KO. 1979. Apathy and death in early Jamestown. Journal of American History 66: 24–40.
———. 2007. Roanoke: The abandoned colony. Lanham, Md.: Rowman and Littlefield.
Kuzeev RG, Mukhamediarov SF. 1992. Ethnolinguistic relationships—the ethnogenesis and ethnocultural relationships of the Turkic peoples of the Volga and Urals—problems and tasks. Anthropology and Archeology of Eurasia 31: 24–39.
Landers J. 1997. Africans in the Spanish colonies. Historical Archaeology 31: 84–91.
Langfur H. 2005. Moved by terror: Frontier violence as cultural exchange in late-colonial Brazil. Ethnohistory 52: 255–289.
Lanning JT. 1935. The Spanish missions of Georgia. Chapel Hill: University of North Carolina Press.
Larsen CS. 1982. The anthropology of St. Catherines Island: 3. Prehistoric human biological adaptation. Anthropological Papers of the American Museum of Natural History, no. 57, pt 3.
———. 1990. The archaeology of mission Santa Catalina de Guale: 2. Biocultural interpretations of a population in transition. Anthropological Papers of the American Museum of Natural History, no. 68.
———. 1993. On the frontier of contact: Mission bioarchaeology in La Florida. In: McEwan BG, ed. The Spanish missions of La Florida. Gainesville: University Press of Florida. 322–356.
———. 1997. Bioarchaeology: Interpreting behavior from the human skeleton. Cambridge: Cambridge University Press.
———. 2001. Bioarchaeology of Spanish Florida: The impact of colonialism. Gainesville: University Press of Florida.
———. 2002. Bioarchaeology of the late prehistoric Guale. South End Mound I, St. Catherines Island, Georgia. Anthropological Papers of the American Museum of Natural History, no. 84.
Larsen CS, Crosby AW, Griffin MC, Hutchinson DL, Ruff CB, Russell KF, Schoeninger MJ, Sering LE, Simpson SW, Takács JL, Teaford MF. 2002. A biohistory of health and behavior in the Georgia Bight: The agricultural transition and the impact of European contact. In: Steckel RH, Rose JC, eds. The backbone of history: Health and nutrition in the Western Hemisphere. New York: Cambridge University Press. 406–439.
Larsen CS, Harn DE. 1994. Health in transition: Disease and nutrition in the Georgia Bight. In: Sobolik KD, ed. Paleonutrition: The diet and health of prehistoric Americans. Occasional Paper no. 22. Carbondale: Center for Archaeological Investigations, Southern Illinois University-Carbondale. 222–234.
Larsen CS, Hutchinson DL. 1992. Dental evidence for physiological disruption: Biocultural interpretations from the eastern Spanish borderlands, U.S.A. Journal of Paleopathology 2: 151–169.
Larsen CS, Hutchinson DL, Schoeninger MJ, Norr L. 2001. Food and stable iso-

topes in La Florida. Diet and nutrition before and after contact. In: Larsen CS, ed. Bioarchaeology of Spanish Florida: The impact of colonialism. Gainesville: University Press of Florida. 52–81.

Larsen CS, Huynh HP, McEwan BG. 1996. Death by gunshot: Biocultural implications of trauma at mission San Luis. International Journal of Osteoarchaeology 6: 42–50.

Larsen CS, Ruff CB. 1994. The stresses of conquest in Spanish Florida: Structural adaptation and change before and after contact. In: Larsen CS, Milner GR, eds. In the wake of contact: Biological responses to conquest. New York: Wiley-Liss. 21–34.

Larsen CS, Ruff CB, Griffin MC. 1996. Implications of changing biomechanical and nutritional environments for activity and lifeway in the eastern Spanish borderlands. In: Baker BJ, Kealhofer L, eds. Bioarchaeology of Native American adaptation in the Spanish borderlands. Gainesville: University Press of Florida. 95–125.

Larsen CS, Ruff CB, Schoeninger MJ, Hutchinson DL. 1992. Population decline and extinction in La Florida. In: Verano JW, Ubelaker DH, eds. Disease and demography in the Americas. Washington, D.C.: Smithsonian Institution Press. 25–39.

Larsen CS, Sering LE. 2000. Inferring iron-deficiency anemia from human skeletal remains: The case of the Georgia Bight. In: Lambert PM, ed. Bioarchaeological studies of life in the age of agriculture: A view from the Southeast. Tuscaloosa: University of Alabama Press. 116–133.

Larsen CS, Shavit R, Griffin MC. 1991. Dental caries evidence for dietary change: An archaeological context. In: Kelley MA, Larsen CS, eds. Advances in dental anthropology. New York: Wiley-Liss. 179–202.

Larsen CS, Tung TA. 2002. Mission San Luis de Apalachee: Final report on the human remains. Report submitted to Bonnie McEwan, Florida Bureau of Archaeological Research, Tallahassee.

Larson LH, Jr. 1969. Aboriginal subsistence technology on the southeastern coastal plain during the late prehistoric period. PhD dissertation, University of Michigan.

———. 1972. Functional considerations of warfare in the Southeast during the Mississippi period. American Antiquity 37: 383–392.

———. 1978. Historic Guale Indians of the Georgia coast and the impact of the Spanish mission effort. In: Milanich JT, Proctor S, eds. Tacachale: Essays on the Indians of Florida and southeastern Georgia during the historic period. Gainesville: University Presses of Florida. 120–140.

———. 1980. Aboriginal subsistence technology on the southeastern coastal plain during the late prehistoric period. Gainesville: University Presses of Florida.

Law R. 1992. Warfare on the West African slave coast, 1650–1850. In: Ferguson RB, Whitehead NL, eds. War in the tribal zone. Expanding states and indigenous warfare. Santa Fe: School of American Research. 103–126.

Lease LR, Harris EF. 2001. Absence of association between body size and deciduous tooth size in American Black children. Dental Anthropology Journal 15: 7–11.

Leonard IA. 1936. The Spanish re-exploration of the Gulf Coast in 1686. Mississippi Valley Historical Review 22: 547–557.

Lesser A. 1961. Social fields and the evolution of society. Southwestern Journal of Anthropology 17: 40–48.

Levine RA, Campbell DT. 1972. Ethnocentrism: Theories of conflict, ethnic attitudes, and group behavior. New York: J. Wiley and Sons.

Lightfoot KG, Martinez A, Schiff AM. 1998. Daily practice and material culture in pluralistic social settings: An archaeological study of culture change and persistence from Fort Ross, California. American Antiquity 63: 199–222.

Lindstrom RW. 2001. Soviet ethnogenetic theory and the interpretation of the past. In: Terrell JE, ed. Archaeology, language, and history. Essays on culture and ethnicity. Westport, Conn.: Bergin and Garvey. 57–78.

Lockwood WG. 1984. Beyond ethnic boundaries: New approaches in the anthropology of ethnicity. Michigan Discussions in Anthropology, vol. 7. Department of Anthropology, University of Michigan.

Lombardi AV. 1975. A factor analysis of morphogenetic fields in the human dentition. American Journal of Physical Anthropology 42: 99–104.

Lorant S. 1946. The New World: The first pictures of America. New York: Duell, Sloan and Pearce.

Loucks LJ. 1979. Political and economic interactions between Spaniards and Indians: Archeological and ethnohistorical perspectives of the mission system in Florida. PhD dissertation, University of Florida.

———. 1993. Spanish-Indian interaction on the Florida missions: The archaeology of Baptizing Spring. In: McEwan BG, ed. The Spanish missions of La Florida. Gainesville: University Press of Florida. 193–216.

Luer GM. 1993. Calusa canals in southwestern Florida: Routes of tribute and exchange. Florida Anthropologist 42: 89–127.

Lund AA. 2005. Hellenist era and Hellenicity: On the ethnogenesis and the ethnicity of ancient Hellenists. Historia-Zeitschrift fur Alte Geschichte 54: 1–17.

Lyon E. 1974. The enterprise of Florida. Florida Historical Quarterly 52: 411–422.

———. 1990. The enterprise of Florida. In: Thomas DH, ed. Columbian consequences, vol. 2: Archaeological and historical perspectives on the Spanish borderlands east. Washington, D.C.: Smithsonian Institution Press. 281–296.

MacEachern S. 2000. Genes, tribes, and African history. Current Anthropology 41: 357–371.

Magoon D, Norr L, Hutchinson DL, Ewen CR. 2001. An analysis of human skeletal materials from the Snow Beach Site (8WA52). Southeastern Archaeology 20: 18–30.

Mahoney MR. 2003. Racial formation and ethnogenesis from below: The Zulu case, 1879–1906. International Journal of African Historical Studies 36: 559–583.

Malone PM. 1991. The skulking way of war: Technology and tactics among the New England Indians. New York: Madison Books.

Manly BFJ. 1994. Multivariate statistical methods: A primer. London: Chapman and Hall.

Manucy A. 1992. The houses of St. Augustine, 1565–1821. Gainesville: University Press of Florida.

Marks J. 2002. What is molecular anthropology? What can it be? Evolutionary Anthropology 11: 131–135.

Marrinan RA. 1991. Archaeological investigations at mission Patale, 1984–1991. Florida Anthropologist 44: 228–254.

———.1993. Archaeological investigations at mission Patale, 1984–1992. In: McEwan BG, ed. The Spanish missions of La Florida. Gainesville: University Press of Florida. 244–294.

Marrinan RA, Halpern JA, Heide GM, Blackmore C. 2000. Recent investigations at the O'Connell mission site (8LE157), Leon County, Florida. Florida Anthropologist 53: 224–249.

Mason CI. 2005. The archaeology of Ocmulgee Old Fields, Macon, Georgia. Tuscaloosa: University of Alabama Press.

Matsumura H, Nishimoto T. 1996. Statistical analysis on kinship of the Nakazuma Jomon people using tooth crown measurements. Zoo-archaeology 6: 1–17.

Matter RA. 1972. The Spanish missions of Florida: The friars versus the governors in the "Golden Age," 1606–1690. PhD dissertation, University of Washington.

———. 1973. Economic basis of the seventeenth-century Florida missions. Florida Historical Quarterly 52: 18–38.

McAlister LN. 1984. Spain and Portugal in the New World, 1492–1700. Minneapolis: University of Minnesota Press.

McEvoy B, Richards M, Forster P, Bradley DG. 2004. The *Longue Durée* of genetic ancestry: Multiple genetic marker systems and Celtic origins on the Atlantic facade of Europe. American Journal of Human Genetics 75: 693–702.

McEwan BG. 1991a. San Luis de Talimali: The archaeology of Spanish-Indian relations at a Florida mission. Historical Archaeology 25: 36–60.

———. 1991b. The archaeology of women in the Spanish New World. Historical Archaeology 25: 33–41.

———. 1992. Archaeology of the Apalachee village at San Luis de Talimali. Florida Bureau of Archaeological Research, Tallahassee, Fla. Florida Archaeological Report 28.

———. 1993a. The Spanish missions of La Florida. Gainesville: University Press of Florida.

———.1993b. Hispanic life on the seventeenth-century Florida frontier. In: McEwan BG, ed. The Spanish missions of La Florida. Gainesville: University Press of Florida. 295–321.

———. 2000a. Indians of the greater Southeast. Gainesville: University Press of Florida.

———. 2000b. The Apalachee Indians of Northwest Florida. In: McEwan BG, ed. Indians of the greater Southeast. Gainesville: University Press of Florida. 57–84.

———. 2001. The spiritual conquest of La Florida. American Anthropologist 103: 633–644.

McEwan BG, Poe CB. 1994. Excavations at Fort San Luis. Florida Anthropologist 47: 90–106.

McGuire RH. 1982. The study of ethnicity in historical archaeology. Journal of Anthropological Archaeology 1: 159–178.

———. 1983. Ethnic group, status and material culture at Rancho Punta de Agua. In: Ward AE, ed. Forgotten places and things: Archaeological perspectives on American history. Albuquerque: Center for Anthropological Studies. 193–203.

McKay J. 1982. An exploratory synthesis of primordial and mobilizationist approaches to ethnic phenomena. Ethnic and Racial Studies 5: 395–420.

McMurray JA. 1973. The definition of the ceramic complex at San Juan del Puerto. MA thesis, University of Florida.

Merritt JD. 1983. Beyond the town walls: The Indian element in colonial St. Augustine. In: Deagan K, ed. Spanish St. Augustine: The archaeology of a colonial Creole community. New York: Academic Press. 125–147.

Michael HN. 1962. Studies in Siberian ethnogenesis. Toronto: University of Toronto Press.

Milanich JT. 1972. Excavations at the Richardson site, Alachua County, Florida: An early 17th century Potano Indian village (with notes on Potano culture change). Bureau of Historic Sites and Properties, Bulletin no. 2. Division of Archives, History, and Records Management, Tallahassee, FL.

———. 1978. The western Timucua: Patterns of acculturation and change. In: Milanich JT, Proctor S, eds. Tacachale: Essays on the Indians of Florida and southeastern Georgia during the historic period. Gainesville: University Presses of Florida. 59–88.

———. 1990. The European entrada in La Florida: An overview. In: Thomas DH, ed. Columbian consequences, vol. 2: Archaeological and historical perspectives on the Spanish borderlands east. Washington, D.C.: Smithsonian Institution Press. 3–30.

———. 1996. The Timucua. Cambridge, U.K.: Blackwell.

———. 1999. Laboring in the fields of the Lord. Spanish mission and southeastern Indians. Washington, D.C.: Smithsonian Institution Press.

———. 2000. The Timucua Indians of northern Florida and southern Georgia. In: McEwan BG, ed. Indians of the greater Southeast. Gainesville: University Press of Florida. 1–25.

———. 2004. Timucua. In: Fogelson RD, ed. Handbook of North American Indians. Southeast. Washington, D.C.: Smithsonian Institution Press. 219–228.

———. 2005. The devil in the details. Archaeology 58: 26–31.

Milanich JT, Cordell AS, Knight VJ, Jr., Kohler TA, Sigler-Lavelle BJ. 1984. McKeithen Weeden Island. The culture of northern Florida, A.D. 200–900. New York: Academic Press.

Milanich JT, Proctor S. 1978. Tacachale: Essays on the Indians of Florida and southeastern Georgia during the historic period. Gainesville: University Presses of Florida.

Milanich JT, Sturtevant WC. 1972. Francisco Pareja's 1613 *confessionario*: A documentary source for Timucuan ethnography. Tallahassee: Division of Archives, History, and Records Management, Florida Department of State.

Miller L. 2000. Roanoke: Solving the mystery of the lost colony. London: Jonathan Cape.
Mitchell J. 2005. Negotiating identity politics: Emerging Roma ethnogenesis in the post socialist states of south eastern and central eastern Europe. Polish Sociological Review 152: 383–395.
Mitchem JM. 1989. The Ruth Smith, Weeki Wachee, and Tatham mounds: Archaeological evidence of early Spanish contact. Florida Anthropologist 42: 317–339.
Moiseyev V. 2001. Origins of Uralic-speaking populations: Craniological evidence. Homo 52: 240–253.
Monahan Driscoll E, Larsen CS. 1994. Mission San Pedro y San Pablo de Patale, Leon County, Florida: Human remains inventory. Bioarchaeology Research Laboratories, University of North Carolina, Chapel Hill.
Moore JH. 1994a. Ethnogenetic theory. National Geographic Research & Exploration 10: 10–23.
———. 1994b. Putting anthropology back together again: The ethnogenetic critique of cladistic theory. American Anthropologist 96: 925–948.
———. 2001. Ethnogenetic patterns in native North America. In: Terrell JE, ed. Archaeology, language, and history. Essays on culture and ethnicity. Westport, Conn.: Bergin and Garvey. 31–56.
Moorrees CFA, Reed RB. 1964. Correlations among crown diameters of human teeth. Archives of Oral Biology 9: 685–697.
Mullins PR, Paynter R. 2000. Representing colonizers: An archaeology of creolization, ethnogenesis, and indigenous material culture among the Haida. Historical Archaeology 34: 73–84.
Nagata JA. 1974. What is a Malay? Situational selection of ethnic identity in a plural society. American Ethnologist 1: 331–350.
Naroll R. 1964. On ethnic unit classification. Current Anthropology 5: 283–291.
North KE, Martin LJ, Crawford MH. 2000. The origins of the Irish Travellers and the genetic structure of Ireland. Annals of Human Biology 27: 453–465.
Nystrom KC. 2006. Late Chachapoya population structure prior to Inka conquest. American Journal of Physical Anthropology 131: 334–342.
Oatis SJ. 2008. A colonial complex: South Carolina's frontiers in the era of the Yamasee war, 1680–1730. Lincoln: University of Nebraska Press.
Oota H, Saitou N, Matsushita T, Ueda S. 1995. A genetic study of 2,000-year-old human remains from Japan using mitochondrial DNA sequences. American Journal of Physical Anthropology 98: 133–145.
Oré LG. 1936. The martyrs of Florida. New York: Joseph F. Wagner.
Orser CE, Jr. 2004. Race and practice in archaeological interpretation. Philadelphia: University of Pennsylvania Press.
Oshanin LV. 1964. Anthropological composition of the population of Central Asia, and the ethnogenesis of its peoples: I. Russian Translation Series of the Peabody Museum of Archaeology and Ethnology, Harvard University, Volume 2, Number 1. Cambridge, Mass.: Peabody Museum.
Ossenberg NS, Dodo Y, Maeda T, Kawakubo Y. 2006. Ethnogenesis and craniofa-

cial change in Japan from the perspective of nonmetric traits. Anthropological Science 114: 99–115.
Otterbein KF. 1999. A history of research on warfare in anthropology. American Anthropologist 101: 794–805.
Otto JS, Lewis RL. 1974. A formal and functional analysis of San Marcos pottery from site SA 16-23 St. Augustine, Florida. Bureau of Historic Sites and Properties, Bulletin no. 4. Tallahassee, Fla.
Pálsson J. 1978. Some anthropological characteristics of Icelanders analyzed with regard to the problem of ethnogenesis. Journal of Human Evolution 7: 695–702.
Parkinson WA. 2002. The archaeology of tribal societies. Archaeological series number 15. Ann Arbor, Mich.: International Monographs in Prehistory.
Parmenter J. 2007. After the mourning wars: The Iroquois as allies in colonial North American campaigns, 1676–1760. William and Mary Quarterly 64: 39–82.
Pauketat TR. 2000. Politicization and community in the Pre-Columbian Mississippi Valley. In: Canuto MA, Yaeger J, eds. The archaeology of communities: A New World perspective. London: Routledge. 16–43.
———. 2001. Practice and history in archaeology: An emerging paradigm. Anthropological Theory 1: 73–98.
———. 2003. Resettled farmers and the making of a Mississippian polity. American Antiquity 68: 39–66.
Perdue T, Green MD. 2001. The Columbia guide to American Indians of the Southeast. New York: Columbia University Press.
Peterson HL. 1956. Arms and armor in colonial America, 1526–1783. New York: Bramhall House.
Piatek BJ. 1985. Non-local aboriginal ceramics from early historic contexts in St. Augustine. Florida Anthropologist 38: 81–89.
Pietrusewsky M. 2000. Metric analysis of skeletal remains: Methods and applications. In: Katzenberg MA, Saunders SR, eds. Biological anthropology of the human skeleton. New York: Wiley-Liss. 375–416.
Pluckhahn TJ, Ethridge R. 2006. Light on the path. The anthropology and history of the southeastern Indians. Tuscaloosa: University of Alabama Press.
Porter KW. 1949. The Founder of the "Seminole Nation" Secoffee or Cowkeeper. Florida Historical Quarterly 27: 362–384.
Posukh OL, Osipova LP, Kashinskaya YO, Ivakin EA, Kryukov YA, Karafet TM, Kazakovtseva MA, Skobel'tsina LM, Crawford MG, Lefranc MP, Lefranc G. 1998. A genetic study of the south Altaian population of the Mendur-Sokkon village, Altai Republic. Genetika 34: 106–113.
Potter RH, Nance WE, Yu P-L, Davis WB. 1976. A twin study of dental dimension II. Independent genetic determinants. American Journal of Physical Anthropology 44: 397–412.
Potter RHY, Rice JP, Dahlberg AA, Dahlberg T. 1983. Dental size traits within families: Path analysis for first molar and lateral incisor. American Journal of Physical Anthropology 61: 283–289.
Powell JF. 1995. Dental variation and biological affinity among middle Holocene human populations in North America. PhD dissertation, Texas A&M University.

Powell JF, Neves WA. 1999. Craniofacial morphology of the first Americans: Pattern and process in the peopling of the New World. Yearbook of Physical Anthropology 42: 153–188.

Praetzellis A, Praetzellis M, Brown M, III. 1987. Artifacts as symbols of identity: An example from Sacramento's gold rush era Chinese community. In: Staski E, ed. Living in cities: Current research in urban archaeology. Special Publication Series, number 5. Society for Historical Archaeology. 38–47.

Pratt ML. 1991. Arts of the contact zone. Profession 91: 33–40.

Priestley HE. 1928. The Luna papers. Documents relating to the expedition of Don Tristán de Luna y Arellano for the conquest of La Florida in 1559–1561. Volume 1. Deland: Florida State Historical Society.

Quinn WW. 1993. Intertribal integration: The ethnological argument in *Duro v. Reina*. Ethnohistory 40: 34–69.

Ramenofsky AF. 1987. Vectors of death: The archaeology of European contact. Albuquerque: University of New Mexico Press.

———. 1991. Loss of innocence: Explanations of differential persistence in the sixteenth-century Southeast. In: Thomas DH, ed. Columbian consequences, vol. 2. Archaeological and historical perspectives on the Spanish borderlands east. Washington, D.C.: Smithsonian Institution Press. 31–48.

Ramenofsky AF, Wilbur AK, Stone AC. 2003. Native American disease history: Past, present and future directions. World Archaeology 35: 241–257.

Reding K. 1935. Plans for the colonization and defense of Apalachee, 1675. Georgia Historical Quarterly 9: 169–175.

Redmond EM. 2001. Review of "War in the Tribal Zone: Expanding States and Indigenous Warfare." Ethnohistory 48: 550–552.

Reilly SE. 1981. A marriage of expedience: The Calusa Indians and their relations with Pedro Menéndez de Avilés in southwest Florida, 1566–1569. Florida Historical Quarterly 59: 395–421.

Relethford JH. 1991a. Genetic drift and anthropometric variation in Ireland. Human Biology 63: 155–165.

———. 1991b. Effects of changes in population size on genetic microdifferentiation. Human Biology 63: 629–641.

———. 1996. Genetic drift can obscure population history: Problem and solution. Human Biology 68: 29–44.

———. 2001. Global analysis of regional differences in craniometric diversity and population substructure. Human Biology 73: 629–636.

———. 2003. Anthropometric data and population history. In: Herring DA, Swedlund AC, eds. Human biologists in the archives: Demography, health, nutrition, and genetics in historical populations. Cambridge: Cambridge University Press. 32–52.

Relethford JH, Blangero J. 1990. Detection of differential gene flow from patterns of quantitative variation. Human Biology 62: 5–25.

Relethford JH, Lees FC. 1982. The use of quantitative traits in the study of human population structure. Yearbook of Physical Anthropology 25: 113–132.

Relethford JH, Lees FC, Crawford MH. 1997. Population structure and anthropometric variation in rural western Ireland: Migration and biological differentiation. Annals of Human Biology 7: 411–428.

Reynolds LT, Lieberman L. 1996. Race and other misadventures. Essays in honor of Ashley Montagu in his ninetieth year. New York: General Hall.

Reynolds V. 1980. Sociobiology and the idea of primordial discrimination. Ethnic and Racial Studies 3: 303–315.

Ricaut FX, Kolodesnikov S, Keyser-Tracqui C, Alekseev AN, Crubézy E, Ludes B. 2006. Molecular genetic analysis of 400-year-old human remains found in two Yakut burial sites. American Journal of Physical Anthropology 129: 55–63.

Rizk OT, Amugongo SK, Mahaney MC, Hlusko LJ. 2008. The quantitative genetic analysis of primate dental variation: History of the approach and prospects for the future. In: Irish JD, Nelson GC, eds. Technique and application in dental anthropology. New York: Cambridge University Press. 317–346.

Romans B. 1961. A concise natural history of east and west Florida. New Orleans: Pelican.

Roosens EE. 1989. Creating ethnicity: The process of ethnogenesis. Newbury Park, Calif.: Sage.

Rountree HC. 1993. The Powhatans and the English: A case of multiple conflicting agendas. In: Rountree HC, ed. Powhatan foreign relations 1500–1722. Charlottesville: University Press of Virginia. 173–205.

———. 2002. Trouble coming southward: Emanations through and from Virginia 1607–1675. In: Ethridge R, Hudson C, eds. The transformation of the southeastern Indians, 1540–1760. Jackson: University Press of Mississippi. 65–78.

Rountree HC, Turner ER III. 1994. On the fringe of the Southeast: The Powhatan paramount chiefdom in Virginia. In: Hudson C, Tesser CC, eds. The forgotten centuries: Indians and Europeans in the American South, 1521–1704. Athens: University of Georgia Press. 355–372.

Rubin PH. 2000. Does ethnic conflict pay? Politics and the Life Sciences 19: 59–68.

Ruff CB, Larsen CS. 1990. Postcranial biomechanical adaptations to subsistence strategy changes on the Georgia coast. In: Larsen CS, ed. The archaeology of mission Santa Catalina de Guale: 2. Biocultural interpretations of a population in transition. New York: Anthropological Papers of the American Museum of Natural History, no. 68. 94–120.

———. 2001. Reconstructing behavior in Spanish Florida: The biomechanical evidence. In: Larsen CS, ed. Bioarchaeology of Spanish Florida: The impact of colonialism. Gainesville: University Press of Florida. 113–145.

Sackett JR. 1982. Approaches to style in lithic archaeology. Journal of Anthropological Archaeology 1: 59–112.

———. 1985. Style and ethnicity in the Kalahari: A reply to Wiessner. American Antiquity 50: 154–159.

Sattler RA. 1987. Seminoli *italwa*: Socio-political change among the Oklahoma Seminoles between removal and allotment, 1836–1905. PhD dissertation, University of Oklahoma.

———.1996. Remnants, renegades, and runaways: Seminole ethnogenesis reconsidered. In: Hill JD, ed. History, power and identity. Ethnogenesis in the Americas, 1492–1992. Iowa City: University of Iowa Press. 36–69.

Saunders R. 1992. Guale Indian pottery: A Georgia legacy in northeast Florida. Florida Anthropologist 45: 139–147.

———.1993. Architecture of the missions Santa María and Santa Catalina de Amelia. In: McEwan BG, ed. The Spanish missions of La Florida. Gainesville: University Press of Florida. 35–61.

———. 2000a. The Guale Indians of the lower Atlantic coast: Change and continuity. In: McEwan BG, ed. Indians of the greater Southeast. Gainesville: University Press of Florida. 26–56.

———. 2000b. Stability and change in Guale Indian pottery A.D. 1300–1702. Tuscaloosa: University of Alabama Press.

Scarry JF. 1985. A proposed revision of the Fort Walton ceramic typology: A type-variety system. Florida Anthropologist 38: 199–233.

———. 1994. The Apalachee chiefdom: A Mississippian society on the fringe of the Mississippian world. In: Hudson C, Tesser CC, eds. The forgotten centuries. Indians and Europeans in the American Southeast 1521–1704. Athens: University of Georgia Press. 156–178.

———.1999. Elite identities in Apalachee Province: The construction of identity and cultural change in a Mississippian society. In: Robb J, ed. Material symbols: Culture and economy in prehistory. Occasional Paper no. 26. Carbondale: Center for Archaeological Investigations, Southern Illinois University-Carbondale. 342–361.

Scarry JF, Maxham MD. 2002. Elite actors in the protohistoric: Elite identities and interaction with Europeans in the Apalachee and Powhatan chiefdoms. In: Wesson CB, Rees MA, eds. Between contacts and colonies. Archaeological perspectives on the protohistoric Southeast. Tuscaloosa: University of Alabama Press. 142–169.

Scherer AK. 2007. Population structure of the classic period Maya. American Journal of Physical Anthropology 132: 367–380.

Schillaci MA. 2003. The development of population diversity at Chaco Canyon. Kiva 68: 221–245.

Schillaci MA, Stojanowski CM. 2005. Craniometric variation and population history of the prehistoric Tewa. American Journal of Physical Anthropology 126: 404–412.

Schnutenhaus S, Rösing FW. 1998. World variation of tooth size. In: Alt KW, Rösing FW, Teschler-Nicola M, eds. Dental anthropology: Fundamentals, limits, and prospects. New York: Springer. 521–535.

Schoeninger MJ, van Der Merwe NJ, Moore K, Lee-Thorp J, Larsen CS. 1990. Decrease in diet quality between the prehistoric and the contact periods. In: Larsen CS, ed. The archaeology of mission Santa Catalina de Guale: 2. Biocultural interpretations of a population in transition. New York: Anthropological Papers of the American Museum of Natural History, no. 68. 78–93.

Schultz M, Larsen CS, Kreutz K. 2001. Disease in Spanish Florida: Microscopy of porotic hyperostosis and cribra orbitalia. In: Larsen CS, ed. Bioarchaeology of Spanish Florida: The impact of colonialism. Gainesville: University Press of Florida. 207–225.

Schwartz GM, Nichols JJ. 2006. After collapse: The regeneration of complex societies. Tucson: University of Arizona Press.

Scott GR, Turner CG. 1997. The anthropology of modern human teeth: Dental morphology and its variation in recent human populations. Cambridge: Cambridge University Press.

Seaberg LM. 1955. The Zetrouer site: Indian and Spanish in central Florida. MA thesis, University of Florida.

———. 1991. Report on the Indian site at the "Fountain of Youth," St. Augustine. In: Deagan KA, ed. America's ancient city, Spanish St. Augustine, 1565–1763. New York: Garland. 209–274.

Sears WH. 1959. Two Weeden Island period burial mounds, Florida. Contributions of the Florida State Museum, No. 5. University of Florida, Gainesville.

———. 1967. The Tierra Verde burial mound. Florida Anthropologist 20: 25–73.

Shapiro G, McEwan BG. 1992. Archaeology at San Luis. Part one: The Apalachee council house. Florida Archaeology, no. 6. Tallahassee: Florida Bureau of Archaeological Research.

Shapiro G, Vernon R. 1992. Archaeology at San Luis. Part two: The church complex. Florida Archaeology, no. 6. Tallahassee: Florida Bureau of Archaeological Research.

Sharma K, Corruccini RS, Henderson AM. 1985. Genetic variance in dental dimensions of Punjabi twins. Journal of Dental Research 64: 1389–1391.

Sharrock SR. 1974. Crees, Cree-Assiniboines, and Assiniboines: Interethnic social organization on the far northern plains. Ethnohistory 21: 95–122.

Shennan SJ. 1989. Introduction: Archaeological approaches to cultural identity. In: Shennan SJ, ed. Archaeological approaches to cultural identity. London: Unwin Hyman. 1–32.

———. 1993. After social evolution: A new archaeological agenda? In: Yoffee N, Sherratt A, eds. Archaeological theory: Who sets the agenda? Cambridge: Cambridge University Press. 53–59.

Shimada I, Shinoda K, Farnum J, Corruccini R, Watanabe H. 2004. An integrated analysis of pre-Hispanic mortuary practices: A Middle Sicán case study. Current Anthropology 45: 369–402.

Shinoda K, Kanai S. 1999. Intracemetery genetic analysis at the Nakazuma Jomon site in Japan by mitochondrial DNA sequencing. Anthropological Science 107: 129–140.

Shinoda K, Kunisada T. 1994. Analysis of ancient Japanese society through mitochondrial DNA sequencing. International Journal of Osteoarchaeology 4: 291–297.

Shinoda K, Matsumura H, Nishimoto T. 1998. Genetical and morphological analysis on kinship of the Nakazuma Jomon people using mitochondrial DNA and tooth crown measurements. Zoo-archaeology 11: 1–21.

Shubin PN, Morokov VA, Efimtseva EA, Chelpanova TI. 1997. Ethnogenetics of

Komi-Zyryans according to distribution of gene frequencies of erythrocyte and serum systems of blood groups. Genetika 33: 235–242.

Sider GM. 1976. Lumbee Indian cultural nationalism and ethnogenesis. Dialectical Anthropology 1: 161–172.

———. 1994. Identity as history. Ethnohistory, ethnogenesis and ethnocide in the southeastern United States. Identities 1: 109–122.

Simpson SW. 2001. Patterns of growth disruption in La Florida: Evidence from enamel microstructure. In: Larsen CS, ed. Bioarchaeology of Spanish Florida: The impact of colonialism. Gainesville: University Press of Florida. 146–180.

Simpson SW, Hutchinson DL, Larsen CS. 1990. Coping with stress: Tooth size, dental defects and age-at-death. In: Larsen CS, ed. The archaeology of Mission Santa Catalina de Guale: 2. Biocultural interpretations of a population in transition. Anthropological Papers of the American Museum of Natural History, no. 68. 66–77.

Sjøvold T. 1984. A report on the heritability of some cranial measurements and nonmetric traits. In: van Vark GN, Howells WW, eds. Multivariate statistical methods in physical anthropology. Dordrecht: D. Reidel. 233–246.

Slezkine Y. 1996. N. Ia. Marr and the national origins of Soviet ethnogenetics. Slavic Review 55: 826–862.

Smith MT. 1987. Archaeology of aboriginal culture change in the interior Southeast: Depopulation during the early historic period. Gainesville: University Presses of Florida.

———. 1994. Archaeological investigations at the Dyar site, 9GE5. Laboratory of Archaeology Series, Report no. 32. Athens: University of Georgia.

Smith MT, Hally DJ. 1992. Chiefly behavior: Evidence from sixteenth century Spanish accounts. Archeological Papers of the American Anthropological Association 3: 99–109.

Sofaer JR. 2006. The body as material culture: A theoretical osteoarchaeology. Cambridge: Cambridge University Press.

Sparks CS, Jantz RL. 2002. A reassessment of human cranial plasticity: Boas revisited. Proceedings of the National Academy of Sciences USA 99: 14636–14639.

Spellman CW. 1965. The "Golden Age" of the Florida mission, 1632–1674. Catholic Historical Review 51: 354–378.

Spencer S. 2006. Race and ethnicity: Culture, identity and representation. London: Routledge.

Stahle DW, Cleaveland MK, Blanton DB, Therrell MD, Gay DA. 1998. The lost colony and Jamestown droughts. Science 280: 564–567.

Staski E. 1987. Border city, border culture: Assimilation and change in late 19th century El Paso. In: Staski E, ed. Living in cities: Current research in urban archaeology. Special Publication Series, number 5. Society for Historical Archaeology. 48–55.

Steadman DW. 1998. The population shuffle in the Central Illinois Valley: A diachronic model of Mississippian biocultural interactions. World Archaeology 30: 306–326.

———. 2001. Mississippians in motion? A population genetic analysis of interregional gene flow in west-central Illinois. American Journal of Physical Anthropology 114: 61–73.
Stefan VH. 1999. Craniometric variation and homogeneity in prehistoric/protohistoric Rapa Nui (Easter Island) regional populations. American Journal of Physical Anthropology 110: 407–419.
Steinen KT. 1992. Ambushes, raids, and palisades: Mississippian warfare in the interior Southeast. Southeastern Archaeology 11: 132–139.
Steinen KT, Ritson R. 1996. In defense of the frontier: Considerations of Apalache warfare during the period 1539–1540. Florida Anthropologist 49: 111–120.
Stick D. 1983. Roanoke Island. The beginnings of English America. Chapel Hill: University of North Carolina Press.
Stojanowski CM. 2001. Cemetery structure, population aggregation and biological variability in the mission centers of La Florida. PhD dissertation, University of New Mexico.
———. 2003a. Matrix decomposition model for investigating prehistoric intracemetery biological variation. American Journal of Physical Anthropology 122: 216–231.
———. 2003b. Differential phenotypic variability among the Apalachee mission populations of La Florida: A diachronic perspective. American Journal of Physical Anthropology 120: 352–363.
———. 2004. Population history of native groups in pre- and postcontact Spanish Florida: Aggregation, gene flow, and genetic drift on the southeastern U.S. Atlantic Coast. American Journal of Physical Anthropology 123: 316–322.
———. 2005a. Biocultural histories in La Florida: A bioarchaeological perspective. Tuscaloosa: University of Alabama Press.
———. 2005b. Spanish colonial effects on Native American mating structure and genetic variability in northern and central Florida: Evidence from Apalachee and western Timucua. American Journal of Physical Anthropology 128: 273–286.
———. 2005c. The bioarchaeology of identity in Spanish colonial Florida: Social and evolutionary transformation before, during, and after demographic collapse. American Anthropologist 107: 417–431.
———. 2005d. Biological structure of the San Pedro y San Pablo de Patale mission cemetery. Southeastern Archaeology 24: 165–179.
Stojanowski CM, Buikstra JE. 2004. Biodistance analysis, a biocultural enterprise: A rejoinder to Armelagos and Van Gerven (2003). American Anthropologist 106: 430–431.
———. 2005. Research trends in human osteology: A content analysis of papers published in the American Journal of Physical Anthropology. American Journal of Physical Anthropology 128: 98–109.
Stojanowski CM, Schillaci MA. 2006. Phenotypic approaches for understanding patterns of intracemetery biological variation. Yearbook of Physical Anthropology 49: 49–88.
Stone T. 2003. Social identity and ethnic interaction in the western pueblos of the American Southwest. Journal of Archaeological Method and Theory 10: 31–66.

Storey R. 1986. Diet and health comparisons between pre- and post-Columbian Native Americans in North Florida. American Journal of Physical Anthropology 69: 268.

Sturtevant WC. 1962. Spanish-Indian relations in southeastern North America. Ethnohistory 9: 41–94.

———. 1971. Creek into Seminole. In: Leacock E, Lurie N, eds. North American Indians in historical perspective. New York: Random House. 92–128.

———.1983. Tribe and state in the sixteenth and twentieth centuries. In: Tooker E, ed. The development of political organization in native North America. Washington, D.C.: American Ethnological Society. 3–16.

Susanne C. 1977. Heritability of anthropological characters. Human Biology 49: 573–580.

Swanton JR. 1922. Early history of the Creek Indians and their neighbors. Washington, D.C.: Smithsonian Institution Press.

———. 1946. The Indians of the southeastern United States. Washington, D.C.: Smithsonian Institution Press.

Symes MI, Stephens ME. 1965. A 272: The Fox Pond site. Florida Anthropologist 18: 65–76.

Taitt D. 1916. Journal of David Taitt's travels from Pensacola, west Florida, to and through the country of the Upper and the Lower Creeks, 1772. In: Mereness ND, ed. Travels in the American colonies. New York: Macmillan. 493–565.

Tartaglia M, Scano G, De Stefano GF. 1996. An anthropogenetic study on the Oromo and Amhara of central Ethiopia. American Journal of Human Biology 8: 505–516.

Tatarek NE, Sciulli PW. 2000. Comparison of population structure in Ohio's Late Archaic and Late Prehistoric periods. American Journal of Physical Anthropology 112: 363–376.

Teaford MF, Larsen CS, Pastor RF, Noble VE. 2001. Pits and scratches: Microscopic evidence of tooth use and masticatory behavior in La Florida. In: Larsen CS, ed. Bioarchaeology of Spanish Florida: The impact of colonialism. Gainesville: University Press of Florida. 82–112.

Tefft SK. 1999. Perspectives on panethnogenesis: The case of the Montagnards. Sociological Spectrum 19: 387–400.

TePaske JJ. 1964. The governorship of Spanish Florida, 1700–1763. Durham, N.C.: Duke University Press.

Terrell JE. 2001a. Archaeology, language and history: Essays on culture and ethnicity. Westport, Conn.: Bergin and Garvey.

———. 2001b. Introduction. In: Terrell JE, ed. Archaeology, language and history: Essays on culture and ethnicity. Westport, Conn.: Bergin and Garvey. 1–10.

———. 2001c. The uncommon sense of race, language, and culture. In: Terrell JE, ed. Archaeology, language and history: Essays on culture and ethnicity. Westport, Conn.: Bergin and Garvey. 11–30.

Terrell JE, Hunt TL, Gosden C. 1997. The dimensions of social life in the Pacific: Human diversity and the myth of the primitive isolate. Current Anthropology 38: 155–195.

Thomas DH. 1988. Saints and soldiers at Santa Catalina: Hispanic designs for colonial America. In: Leone MP, Potter PB, eds. The recovery of meaning: Historical archaeology in the eastern United States. New York: Percheron Press. 73–140.

———. 1990a. The Spanish missions of La Florida: An overview. In: Thomas DH, ed. Columbian consequences, vol. 2. Archaeological and historical perspectives on the Spanish borderlands east. Washington, D.C.: Smithsonian Institution Press. 357–397.

———. 1990b. Introduction. In: Larsen CS, ed. The archaeology of mission Santa Catalina de Guale: 2. Biocultural interpretations of a population in transition. Anthropological Papers of the American Museum of Natural History, no. 68. 8–10.

Thomas DH, Larsen CS. 1979. The anthropology of St. Catherines Island: 2. The Refuge-Deptford mortuary complex. Anthropological Papers of the American Museum of Natural History 56, part 1.

Thunen RL, Ashley KH. 1995. Mortuary behavior along the lower St. Johns: An overview. Florida Anthropologist 48: 3–12.

Townsend GC, Brown T. 1978a. Heritability of permanent tooth size. American Journal of Physical Anthropology 49: 497–504.

———. 1978b. Inheritance of tooth size in Australian Aboriginals. American Journal of Physical Anthropology 48: 305–314.

———. 1979. Family studies of tooth size factors in the permanent dentition. American Journal of Physical Anthropology 50: 183–190.

Townsend GC, Brown T, Richards LC, Rogers JR, Pinkerton SK, Travan GR, Burgess VB. 1986. Metric analyses of the teeth and faces of South Australian twins. Acta Geneticae Medicae et Gemellologiae 35: 179–192.

Townsend G, Richards L, Hughes T. 2003. Molar intercuspal dimensions: Genetic input to phenotypic variation. Journal of Dental Research 82: 350–355.

Trosper RL. 1981. American Indian nationalism and frontier expansion. In: Keyes CF, ed. Ethnic change. Seattle: University of Washington Press. 246–270.

Trottier RW. 1981. Charters of panethnic identity: Indigenous American Indians and immigrant Asian-Americans. In: Keyes CF, ed. Ethnic change. Seattle: University of Washington Press. 272–305.

True DO. 1945. Memoire of d'Escalante Fontaneda respecting Florida. Coral Gables, Fla.: Glade House.

Turner VW. 1969. The ritual process: Structure and anti-structure. Chicago: Aldine.

Ubelaker DH. 1978. Human skeletal remains: Excavation, analysis, interpretation. Washington, D.C.: Taraxacum.

Udina IG, Rautian GS. 1994. Genetic polymorphism of the HLA system in Komi and Komi-Permyaks. Genetika 30: 982–991.

Udina IG, Rychkov YG, Malenko AF, Shkurko VN, Khodjaev EY, Donyeva TV. 1985. Study of the HLA system of the Uzbek population of the Ferghana Valley: The HLA antigens, genes and haplotypes of the Uzbek population in regard to its ethnogenesis. Genetika 21: 161–167.

Upton D. 1996. Ethnicity, authenticity, and invented traditions. Historical Archaeology 30: 1–7.
Vail L. 1989. The creation of tribalism in southern Africa. Berkeley: University of California Press.
van den Berghe P. 1978. Race and ethnicity: A sociobiological perspective. Ethnic and Racial Studies 1: 401–411.
———. 1981. The ethnic phenomenon. New York: Elsevier.
———. 1986. Ethnicity and the sociobiology debate. In: Rex J, Mason D, eds. Theories of race and ethnic relations. Cambridge: Cambridge University Press. 246–263.
van Gennep A. 1960. The rites of passage. Chicago: University of Chicago Press.
Varela HH, Cocilovo JA. 2002. Genetic drift and gene flow in a prehistoric population of the Azapa valley and coast, Chile. American Journal of Physical Anthropology 118: 259–267.
Varner JG, Varner J. 1951. The Florida of the Inca. Austin: University of Texas Press.
Voss BL. 2005. From *casta* to *Californio*: Social identity and the archaeology of culture contact. American Anthropologist 107: 461–474.
Wallace RL. 1975. An archeological, ethnohistoric, and biochemical investigation of the Guale aborigines of the Georgia coastal strand. PhD dissertation, University of Florida.
Wallerstein I. 2004. World-systems analysis: An introduction. Durham, N.C.: Duke University Press.
Waselkov GA. 1989. Seventeenth-century trade in the colonial Southeast. Southeastern Archaeology 8: 117–133.
Waselkov GA, Cottier JW. 1984. European perceptions of eastern Muskogean ethnicity. In: Boucher P, ed. Proceedings of the tenth meeting of the French Colonial Historical Society, University of Alabama. Lanham, Md.: University Press of America. 23–45.
Wauchope R. 1966. Archaeological survey of northern Georgia with a test of some cultural hypotheses. Memoirs of the Society for American Archaeology, no. 21. Salt Lake City.
Weisman BR. 1989. Like beads on a string. A culture history of the Seminole Indians in northern peninsular Florida. Tuscaloosa: University of Alabama Press.
———. 1992. Excavations on the Franciscan frontier: Archaeology at the Fig Springs mission. Gainesville: University Press of Florida.
———. 1993. Archaeology of Fig Springs mission, Ichetucknee Springs State Park. In: McEwan BG, ed. The Spanish missions of La Florida. Gainesville: University Press of Florida. 165–192.
———. 2000. Archaeological perspectives on Florida Seminole ethnogenesis. In: McEwan BG, ed. Indians of the greater Southeast. Gainesville: University Press of Florida. 299–317.
Wenhold LL. 1936. A 17th century letter of Gabriel Díaz Vara Calderón, bishop of Cuba, describing the Indians and Indian missions of Florida. Smithsonian

Miscellaneous Collection, vol. 95, no. 16. Washington, D.C.: Smithsonian Institution Press. 1–14.

Whitehead NL. 1990. Carib ethnic soldiering in Venezuela, the Guianas, and the Antilles, 1492–1820. Ethnohistory 37: 357–385.

———. 1992. Tribes make states and states make tribes: Warfare and the creation of colonial tribes and states in northeastern South America. In: Ferguson RB, Whitehead NL, eds. War in the tribal zone. Expanding states and indigenous warfare. Santa Fe: School of American Research. 127–150.

Wickman PR. 1999. The tree that bends: Discourse, power and the survival of the Maskókî peoples. Tuscaloosa: University of Alabama Press.

Widmer RJ. 1994. The structure of southeastern chiefdoms. In: Hudson C, Tesser CC, eds. The forgotten centuries. Indians and Europeans in the American South, 1521–1704. Athens: University of Georgia Press. 125–155.

Wierciński A, Bielicki T. 1962. The racial analysis of human populations in relation to their ethnogenesis. Current Anthropology 3: 2,9–46.

Wiessner P. 1983. Style and social information in Kalahari San projectile points. American Antiquity 48: 253–276.

———. 1984. Reconsidering the behavioural basis for style: A case study among the Kalahari San. Journal of Anthropological Archaeology 3: 190–234.

———. 1985. Style or isochrestic variation? A reply to Sackett. American Antiquity 50: 160–166.

———. 1990. Is there a unity to style? In: Conkey M, Hastorf C, eds. Uses of style in archaeology. Cambridge: Cambridge University Press. 105–112.

Williams FL, Belcher RL, Armelagos GJ. 2005. Forensic misclassification of ancient Nubian crania: Implications for assumptions about human variation. Current Anthropology 46: 340–346.

Williams JM. 1983. The Joe Bell site: Seventeenth century lifeways on the Oconee River. PhD dissertation, University of Georgia.

Williams-Blangero S. 1989a. Phenotypic consequences of nonrandom migration in the Jirels of Nepal. American Journal of Physical Anthropology 80: 115–125.

———. 1989b. Clan-structured migration and phenotypic differentiation in the Jirels of Nepal. Human Biology 61: 143–157.

Williams-Blangero S, Blangero J. 1989. Anthropometric variation and the genetic structure of the Jirels of Nepal. Human Biology 61: 1–12.

Wilson RL. 1965. Excavations at the Mayport Mound, Florida. Contributions of the Florida State Museum 13. Gainesville: Florida State Museum.

Wobst HM. 1977. Stylistic behavior and information exchange. In: Cleland CE, ed. For the director: Research essays in honor of James B. Griffin. Ann Arbor: University of Michigan Museum of Anthropology. 317–342.

Wolf ER. 1982. Europe and the people without history. Berkeley: University of California Press.

Wood JW, Milner GR, Harpending HC, Weiss KM. 1992. The osteological paradox: Problems of inferring prehistoric health from skeletal samples. Current Anthropology 33: 343–370.

Worcester DE, Schilz TF. 1984. The spread of firearms among the Indians of the Anglo-French frontiers. American Indian Quarterly 8: 103–115.

Worth JE. 1992. The Timucuan missions of Spanish Florida and the rebellion of 1656. PhD dissertation, University of Florida.

———. 1995. The struggle for the Georgia coast: An eighteenth-century Spanish retrospective on Guale and Mocama. American Museum of Natural History, Anthropological papers, no. 75.

———. 1998a. The Timucuan chiefdoms of Spanish Florida, vol. 1: Assimilation. Gainesville: University Press of Florida.

———. 1998b. The Timucuan chiefdoms of Spanish Florida, vol. 2: Resistance and destruction. Gainesville: University Press of Florida.

———. 2002. Spanish missions and the persistence of chiefly power. In: Ethridge R, Hudson C, eds. The transformation of the southeastern Indians, 1540–1760. Jackson: University Press of Mississippi. 39–64.

———. 2004. Guale. In: Fogelson RD, ed. Handbook of North American Indians. Southeast. Washington, D.C.: Smithsonian Institution Press. 238–244.

———. 2006. Bridging prehistory and history in the Southeast: Evaluating the utility of the acculturation concept. In: Pluckhahn TJ, Ethridge R, eds. Light on the path. The anthropology and history of the southeastern Indians. Tuscaloosa: University of Alabama Press. 196–206.

Wright JL. 1981. The only land they knew: The tragic story of the American Indians in the Old South. New York: Free Press.

———. 1986. Creeks and Seminoles. Lincoln: University of Nebraska Press.

Yaeger J, Canuto MA. 2000. Introducing an archaeology of communities. In: Canuto MA, Yaeger J, eds. Archaeology of communities: A New World perspective. London: Routledge. 1–15.

Yao YG, Lü XM, Luo HR, Li WH, Zhang YP. 2000. Gene admixture in the Silk Road region of China: Evidence from mtDNA and melanocortin 1 receptor polymorphism. Genes and Genetic Systems 75: 173–178.

Yelvington K. 1991. Ethnicity as practice? A comment on Bentley. Comparative Studies in Society and History 33: 158–168.

Zakrzewski SR. 2007. Population continuity or population change: Formation of the ancient Egyptian state. American Journal of Physical Anthropology 132: 501–509.

Index

8Ci203. *See* Tatham Mound
8Co17. *See* McKeithen Mound C
8Du62. *See* Browne Mound
8Du66. *See* Holy Spirit site
8Du96. *See* Mayport Mound
8He12. *See* Weeki Wachee
8Ja65. *See* Waddells Mill Pond
8Le1. *See* Lake Jackson Mounds
8Le170. *See* Killearn Borrow Pit
8Pi51. *See* Tierra Verde Mound
8Wa52. *See* Snow Beach
9Br2. *See* Leake site
9Bry6. *See* Seven Mile Bend
9Ch1. *See* Irene Mound
9Fl5. *See* King site
9G35. *See* Dyar site
9Gn1. *See* Airport site
9Gn51. *See* Kent Mound
9Li18. *See* Johns Mound
9Li20. *See* Marys Mound
9Li26. *See* Seaside Mounds
9Li3. *See* South End Mound
9Li47. *See* McLeod Mound
9Li62. *See* Seaside Mounds
9Mc184. *See* Norman Mound
9Mcl88. *See* Lewis Creek Mounds
9Mg28. *See* Joe Bell site
9Mu100. *See* Sixtoe Field
9Mu101. *See* Bell Field
9Mu102. *See* Little Egypt

9Pm137. *See* Woodlief site
9Tp64. *See* Avery site

Abeika, 146
acculturation, 53, 61
Acuera, 185n7
additive genetic variation, 7, 19, 22, 185n11
adornment, 118, 126
Adrian, chief of Bacuqua, 140, 142
ad sancto, 118. *See also* mortuary practices
Africa, 63. *See also* West Africa
Africans, 51, 60, 87, 93, 98, 136, 137, 189n4, 190n10. *See also* maroons
age effects, 28
age identity, 47, 48, 107
Aguacaleyquen, 185n7
Agua Dulce, 185n7, 188n7
Agua Salada, 185n8
aindiado, 82
Airport site, 159, 163
Ais, 71
Alabamas, 146
Alachua Seminole, 133, 138, 139, 140, 143, 145, 148, 151, 195n4, 195n7
alcohol, 132, 196n10
Alicamani, 185n7
allele, 18, 20, 40, 174
allocation, 7, 55, 175

233

Altamaha River, 13, 28, 113, 160, 184n3
Amacano, 97, 148, 193n19
Amelia Island. *See* Santa María Island
American Revolution, 135, 136
Amhara, 58
ancestor-descendant relationships, 34, 35, 57
ancient DNA, 6
anthropometry, -ics, 29, 35
Apalachee, 2, 7, 11, 17, 133, 145, 186n12; 1647 revolt, 82, 95, 194n6; analytical results for, 30, 34, 35, 41, 42, 165, 166, 168, 170; burial practices, 118; ceramics, 113, 194n3; elite identity, 60; emigration to, 71, 84, 148, 193n19; exodus, 156, 179; firearms, 191n14, 192n15; health, 120, 122, 123, 125, 194n5; as an identity, 1, 66, 81, 99, 101, 108, 128, 130, 144, 150, 172; kinship among chiefs, 72; migration, 62; population size, 123, 194n5; post 1700, 61, 131, 133, 137, 140, 141, 142, 143, 147, 155, 180; province, 24, 35, 87, 148, 158, 189n2; as a pull for proto-Seminole, 138, 139, 142, 148, 153; raids on missions, 83, 88, 90, 96, 97; refugees in, 91, 148, 149, 196n12; regional political standing, 70, 75; response to de Soto, 73; skeletal samples, 24, 25, 26, 27, 28, 29, 118, 159, 160, 161, 165, 177; subsistence, 13, 108, 110; warfare, 65, 75, 132, 188n7, 190n12, 195n2
Apalachee-Havana trade, 189n2
Apalachee Seminole, 134, 138, 139, 140, 145, 148, 151, 195n4, 196n10
Apalachicola (people), 16, 84, 88, 89, 96, 97, 142, 143, 147, 195n2. *See also* Creek; Lower Creek; Uchise
Apalachicola River, 132, 140
Apalachicola Seminole, 148
Arapaha, 185n7
Arawakan, 185n9
Arbuthnot, Alexander, 134
architecture, 3, 53, 54, 56, 110, 173

arquebus, 192n17. *See also* firearms
arthritis, 18
Assiniboin, 51
Astina, 185n7
attrition, dental, 28, 29, 161
Aucilla River, 13
Augustín Báez, Domingo, 184n5
Avery site, 160, 164, 165, 166
Ávila, Father, 118
Aztecs, 13

Bacuqua, 140, 142, 143
bajo campana, 18, 110
Baptizing Spring, 113, 191n14
Barbados, 86, 190n8
Beaver Wars, 87
Bell Field, 160, 165, 166
Biedma, Luys Hernandez de, 70, 72, 188n4
billiard ball model, 4, 171
Biloxi, 189n2
biodistance analysis, 5, 6, 23, 46, 59, 150, 157, 173, 176, 181
biohistorical narrative, 11, 129, 132
biological determinism, 56, 57
biosocial approaches, 57
Black Seminole, 136
Bourdieu, Pierre, 3, 49, 104, 105, 194n1
Bowles, William Augustus, 134, 135, 195n6
Brazil, 190n8
Brims, Emperor 133, 137, 140, 141, 142, 148, 196n8, 196n10
Browne Mound, 160, 164, 166
buccolingual, 8, 9, 28, 29, 162, 163, 165
buffer zones, 3, 70, 176
built landscape, 110, 126
burial practices. *See* mortuary practices
burial shrouds, 118

Cabeza de Vaca, Álvar Núñez, 70, 71, 75, 188n7
cacique, 71, 72, 73, 95, 142
Calanay, 185n7

Calderón, Bishop Gabriel Díaz Vara, 27, 89, 110
Californio, 53, 54
Calos 74. *See also* Carlos
Calusa, 69, 70, 71, 72, 73, 74, 128, 155, 188n5
camino real, 156, 193n19
campo santo, 109, 115
canals, 69
cannibalism, 89
capital (*sensu* Bourdieu), 105, 106, 107. *See also* Bourdieu, Pierre; field; habitus
capitalism, 62
Caravay, 185n7
carbon isotopes, 16
Caribbean, 86, 109, 188n1, 189n3, 189n5, 190n10
caries, 16, 122
Carlos (Calusa chief), 72, 73, 188n1, 188n5. *See also* Calos
Carolina, 85, 88, 90, 93, 97, 98, 129, 133. *See also* South Carolina
Cascange, 185n7
Casqui, 72, 74, 188n6
Casti, 185n7
Castillo de San Marcos, 17, 89, 95, 96
Cautio, 188n1
ceramics, 53, 120, 126, 168
ceramic style: Alachua, 112, 113; in eastern Timucua, 111; Fort Walton, 113; in Guale, 111; Irene phase, 184n3; Leon-Jefferson, 112, 113, 114; paddle-stamped, 112, 114; San Marcos, 111, 112, 113; in St. Augustine 111, 112; St. Johns, 111, 112, 113; transition, 112, 194n3
Chacato, 82, 84, 88, 91, 95, 97, 148, 190n12, 193n19
chaîne opératoire, 55
Chalaque, 99
Charlesfort, 17, 189n2
Charles Town, 17, 84, 86, 90, 95, 132, 189n4

charnel structure, 26, 118
Chattahoochee River, 84, 96, 133, 137, 139, 149
Cherokee, 84, 94, 133, 137, 142
Chesapeake Bay, 54
Cheyenne, 51
Chiahas, 147
Chichimecos, 89
Chickasaw, 94
chiefdom, 13, 14, 37, 71, 72, 78, 98, 103, 132, 171, 184n2, 184n3, 185n7, 188n5, 188n7; boundaries, 66, 70, 77; collapse, 62, 69; cycling, 69, 71; organization, 68, 100, 176; warfare among, 73, 74, 75, 76
Chilili, 185n7
Chilokilichi, 140, 142
Chiluque, 87, 95, 99, 190n11
Chines, 148, 193n19
Chisca, 84, 190n12, 196n13; origin of, 194n21; raids on missions, 82, 83, 88, 95, 97; retaliation against, 192n15
Chislacasliche, 142. *See also* Chilokilichi
Choctaw, 94, 133
Cholupaha, 185n7
cimarrones, 149, 156, 157, 195n4
cladistic, 4, 174
class, 49, 75, 102, 147, 194n11; as an identity, 47, 48, 63, 82, 173
cocina, 110
coffin, 27, 188
Cofitachequi, 70, 73
Colone, 195n2
Columbian Exchange, 120. *See also* demographic collapse; epidemics; pathology; population size, decline in
communitas, 108, 187n4
conquista a fuego y sangre, 184n1
conquista de almas, 184n1
convento, 110, 191n14
Convento de San Francisco, 111, 190n14
Coosa, 73, 194

Index   **235**

Copperbelt of Africa, 63
cosmogony, 44, 130, 151, 154, 156, 172. *See also* worldview
cotton farming, 190n8
Coweta (people), 87, 142, 147
Coweta (town), 132, 133, 140, 141, 146, 195n2
Coweta Resolution, 196n8
Cowkeeper, 138, 140
Coya, 185n7
cranial modification, 8, 55, 181
cranial nonmetric 7
craniometrics, 5, 7; narrow sense heritability of, 8, 29
Cree, 51
Creek, 2, 10, 11, 72, 94, 128, 129, 141, 146, 154, 157, 179, 184n5; Delgado expedition to, 192n15; differentiation from Seminole, 135, 136, 145, 172, 195n6, 195n7; emigration from Georgia, 112, 113, 137; English relations, 133, 137, 196n8; ethnic composition, 194n1; firearm use, 196n9; heartland, 149, 170, 171, 192n14; internal factionalism, 147, 149, 150; origins of, 94, 130, 153, 169; politics, 134, 146, 147, 148, 171, 196n10; raids on missions, 16, 87, 91, 96, 99, 127, 144; reprisals against, 96, 132; skeletal samples, 159, 160, 161. *See also* Lower Creek; proto-Creek
Creek Confederacy, 128, 195n7, 196n8
Creek War of 1814, 134, 135
cribra orbitalia, 17, 121, 124, 125
cross-sectional geometry, 5
Cussita, 146. *See also* Kasihta

Darwinian, 57, 174. *See also* fitness
de Ayllón, Lucas Vázquez, 17, 190n5
debt, 132, 133, 135
decentralization, 62, 63, 69
deerskin trade, 87
de Figueroa, Vasco Porcallo, 190n5
de Junco, Juan (translator), 185n10
de la Vega, Garcilaso, 73, 74, 188n6

de León, Ponce, 13, 17, 188n1, 189n5
de Leturiondo, Alonso, 192n15
Delgado, Marcos, 96, 192n15
de Luna, Tristán, 73
demographic collapse, xiii, 2, 12, 16, 22, 26, 39, 42, 62, 64, 65, 68, 76, 77, 82, 84, 94, 97, 102, 108, 112, 113, 123, 126, 156, 157, 177, 183n1. *See also* population size, decline in
demographic restructuring, 63, 82. *See also* population aggregation
dental attrition. *See* attrition, dental
dental modification, 55, 181
dental morphology, 7; narrow sense heritability of, 29
dental wear. *See* attrition, dental
de Quexos, Pedro, 189n5
de Soto, Hernando, 1, 13, 17, 69, 70, 71, 72, 73, 132, 183, 188n2, 188n6, 188n7
deus ex machina, 105
diaspora, 1, 46, 94, 149, 150
diet, 4, 8, 16, 17, 53, 54, 120, 125, 126, 179
disease, 4, 5, 9, 51, 120, 125, 130, 145, 156, 157, 168, 172, 185n7. *See also* epidemics; pathology
distance statistics, 7, 19, 35, 165, 167, 168
divide and rule, 133
doctrina, 14, 26, 82, 83, 110, 115
dominance (genetics), 19
Domingo Augustín Báez, 184n5
Drake, Sir Francis, 95, 189n3
dress, 107, 126, 173, 179
drift. *See* genetic drift
dual ethnic moiety, 147, 148
Dyar site, 161, 166, 168

earth-fast construction, 54
Edelano, 185n7
Edisto Island, 184n3
effective population size. *See* population size, effective
El Presidio de San Francisco, 53
Elvas, Fidalgo of, 70, 188n6, 190n5
emblemic, 54

emic, 52, 58, 61, 79, 80, 81, 99, 101, 103, 126, 136, 147, 148, 178, 187n4
enamel hypoplasia. *See* linear enamel hypoplasia
encomienda, 85
Enecape, 185n7
enslavement, 51, 64, 74, 89, 99, 127, 129, 133, 150. *See also* slaves
entrada, 1, 13, 17, 71, 73, 132, 180, 188n7, 189n4
epidemics, 9, 16, 51, 77, 85, 125, 127, 130, 132, 145, 156, 157, 168, 172, 176, 177, 179, 180; 1696 smallpox, 149, 195n3; in Apalachee, 123, 194n5; and the slave trade, 190n10; in Timucua, 185n7
epistasis (genetic), 19
Erie, 87
Escamaçu, 75, 87, 184n3, 193n19
Española, 189n5
ethnic amalgamation, 43, 50, 97, 180. *See also* ethnogenesis
ethnic boundary, 47, 150, 178; and Barth, 48, 64; as the focus of analysis, 55, 91; formation, 63, 67, 92; maintenance, 61; permeability, 52, 59, 101; transition, 50, 79, 92, 93
ethnic consciousness, 2, 10, 45, 49, 50, 55, 61, 65, 67, 78, 102
ethnic factionalism, 52, 65, 66, 77, 101
ethnic group(s), 10, 53, 79, 98, 178; composition of the Creek and Seminole, 142, 145, 146, 147, 149, 170; content of, 103, 104, 105, 107; definitions of, 47, 48, 49, 50, 55, 64, 67, 69, 187n4; diversity of in La Florida missions, 91, 126, 193n19; formation of, 43, 44, 52; origins of, 186n2; in the southeastern US, 27, 65, 82, 140, 150; as structural chiefdoms, 69, 177; versus biological populations, 56, 57, 58, 59, 174; versus tribe, 60, 63
ethnic identity, 2, 4, 10, 46, 61, 63, 81, 94, 151, 156, 174, 175, 180; and categorization, 80, 100; and ceramic style, 111, 114; among the Creek, 141, 146, 147; definitions of, 59, 64, 67, 79, 107, 187n4; depth of research on, 48, 53, 186n1; and human biology, 52, 56, 57, 77; origins of, 186n2; and practice theory, 101, 103, 104, 105, 106, 108, 119, 120, 127, 194n1; and primordialism, 102; of specific missions, 196n12; theoretical debate about, 48, 49, 50, 57; transformation, 40, 131; versus community, 187n4; versus race, 56, 58
ethnicity. *See* ethnic identity
ethnic marker, 1
ethnic rivalry, 67, 146, 149
ethnic soldier, 64, 87, 88, 97, 98, 151, 153
ethnic symbolism, 55, 103, 115, 126, 173
ethnocentrism, 129
ethnocide, 64, 129
ethnogenesis, 37, 42, 47, 49, 99; in Africa, 50; American interest in, 186n3; archaeological models for, 54, 55, 56; among colonial Native Americans, 9, 10, 46, 51, 65, 94; and conflict, 64, 66, 91, 97, 102; definition of, 48, 129; failed, 100; and human biology, 51, 52, 57, 58; lack of research on, 186n1; in La Florida, 60, 62, 99, 129, 148, 150, 175, 178; life cycle model, 43, 179; liminal phase, 43, 44, 45, 77, 78, 79, 80, 85, 127, 178; and material culture, 114; models for, 50, 55, 174; as a nonreplicable process, 50; objective elements of, 80, 81, 100, 101; as population origins, 52, 53, 174; and practice theory, 104, 106; and primordialism, 103; reintegration phase, 45; and scale of interaction, 92, 98; separation phase, 45, 66, 68, 71, 76, 77, 80, 177; specific examples of, 53, 97, 187n6; subjective elements of, 81, 102; versus ethnogony, 186n2. *See also* Seminole, ethnogenesis of

Index **237**

ethnogenetic critique, 155
ethnogenetic theory, xiii, 3, 10, 61, 114, 153, 175
ethnogenetic transformation, 3, 10, 40, 42, 50, 52, 76, 99, 101, 108, 110, 127, 131, 148, 171, 178. *See also* identity transformation
ethnogony, 186n1
ethnographers, 61, 63
ethnographic present, 71
ethnography, -ic, 3, 4, 9, 10, 40, 44, 46, 47, 51, 52, 56, 63, 64, 70, 118
ethnonym, 45, 46, 61, 63, 81, 83, 127, 129, 130, 133, 155, 171, 180, 187n4
etic, 61, 79, 98, 100, 101, 136, 151, 187n4
evolutionary archaeology, 55
exogamy, 70, 154

fictive kinship, 63, 72, 187n4
field (*sensu* Bourdieu), 105, 106, 107. *See also* Bourdieu, Pierre; *habitus*
Fig Springs. *See* San Martín de Timucua
firearms, 87, 89, 155, 178, 191n14, 196n9; English provisioning of, 93, 98; illegal trade in, 90; Spanish control of, 86; in Spanish Florida, 190n13, 190n14, 192n15, 192n16, 193n17
First Seminole War, 135
First Spanish Period, 12
fitness, 42, 57
flintlock, 90, 192n17, 193n18. *See also* firearms
Flint River, 134, 139, 140
Fontaneda, Hernando D'Escalante, 71, 188n1, 188n3, 189n5
forensics, 7
Fort Caroline, 17, 189n2
Fort San Marcos, 142
Fountain of Youth site, 191n14
Fox Pond site, 113, 191n14
Freshwater district, 112, 185n7
F_{ST}, 157, 165; calculation of, 29; comparative estimates of, 36, 37; estimates among La Florida samples, 30, 31, 32, 33, 34, 35, 41, 176, 177; interpretation of, 19, 20, 21, 22, 23, 24
fugitives, 16, 77, 82, 125, 127, 141, 148, 149, 152, 153, 156, 157, 171, 179, 193n19
funerary practices. *See* mortuary practices
fur trade, 86, 190n6

Gainesville, 138, 145
gender identity, 47, 48, 107
gene flow, 8, 18, 39, 131, 170, 171, 175; extra-local, 19, 66, 177; and identity, 37, 38, 42, 51, 52, 57, 58, 59, 64, 174, 176; interpretation of, 20, 40; microevolutionary effects of, 9; in Spanish Florida, 143, 144, 149, 152, 164. *See also* migration
generational amnesia, 109
genetic distance, 6, 40, 41, 162, 175; calculation of, 29; estimates among La Florida samples, 29, 30, 31, 32, 33; interpretation of, 20, 21, 22, 23
genetic drift, 8, 9; corrections for, 19, 22, 35; interpretation of, 41, 45, 175, 176; microevolutionary effects of, 20, 39, 40; and population size, 39, 42, 183n1
genocide, 64
genotype-environment interaction, 19, 181
gente de razón, 54
Gilmer Bennett, 1
Goose Creek men, 88
Goudillo, Francisco, 189n5
Gourgues, Dominique de, 72
Grammont, Michel de, 95
grave goods, 118
Great Lakes, 98, 178
group identity, 43, 49, 50, 60, 61, 105, 144. *See also* ethnic identity; national identity; religious identity; tribal identity
gruel, 54

Guadalquini, 95, 185n7
Guale, 142, 145; analytical results for, 29, 30, 31, 32, 33, 34, 35, 41, 42, 163, 164, 165, 168, 170; attacks directed against, 17, 83, 87, 88, 90, 96, 97; burial practices, 118; ceramics, 111, 112; chiefdom boundaries, 184n3; establishing missions among, 13; etymology, 189n1; firearms, 192n15, 192n16; health, 122, 123, 124, 125; as an identity, 61, 65, 66, 81, 99, 101, 108, 128, 130, 144, 148, 150, 172; kinship among chiefs, 72; language, 14, 185n5, 185n10; political structure, 184n2; as proto-Seminole, 11, 131, 147, 155, 156; province, 35, 170; refugees in, 84, 91, 193n19, 196n12; revolts, 95; settlement organization, 111; skeletal samples, 24, 25, 26, 27, 28, 159, 160, 161, 165, 168, 177; warfare, 75, 188n7. *See also* Juanillo revolt
habitus, 10, 49, 80, 104, 105, 106, 107, 194n1
health, xiii, 4, 8, 114; decline in, 16, 17, 18, 122; in Florida missions, 121, 122, 123, 124, 125; and identity, 119, 120, 126, 179
hiliswa, 143
Hitchiti, 140, 142, 144, 146, 155
Hobbesian, 51
Holy Spirit site, 160, 164, 166
Houma, 51
Howells, W.W., 18
Huguenots, 13
Huyache, 87, 95
hypoplastic defects. *See* linear enamel hypoplasia

Ibi, 185n7
Ibihica, 185n7
Icafui, 185n7
iconography, 108
identifiability of cause, 39
identity discourse, 4, 9, 38, 51, 55, 111, 181
identity transformation, 4, 40, 59, 60, 63, 80, 126, 129, 130, 174, 181. *See also* ethnogenetic transformation
ideology, 79, 85, 91, 103, 108, 126, 179; conversion, 13, 68, 110, 115, 126, 179; as an identity, 43, 45, 48, 50, 79, 80, 111; and nativism, 134
immigration, 112, 113, 193n19; of the Westo, 87
inbreeding, 22
Indians of La Costa, 185n7
indios de pueblo, 82
indirect rule, 51
infectious disease, 64. *See also* epidemics
instrumentalism, 49, 51, 62, 101, 102, 103, 187n5
integration: biological, 4, 11, 20, 35, 37, 38, 41, 45, 64, 65, 68, 69, 70, 73, 76, 77, 152, 153, 154, 163, 164, 166, 167, 168, 169, 170, 172, 176, 177; community, 12, 45, 48, 67, 68, 71, 172, 175, 188n1, 188n4; intertribal, 4, 71, 165; political, 71, 177
interbreeding, 174, 175
intermarriage, 43, 52, 56, 58, 144, 146, 150; and political integration, 71, 75
Irene Mound, 28, 118, 119, 160, 166
Iroquois, 87
isolation-by-distance, 30, 31, 32, 41, 42, 70, 163, 165, 170. *See also* population structure

Jackson, Andrew, 134, 135
Jamaica, 95, 190n8
James River, 87
Jamestown, 78, 85, 178, 190n8
Jeaga, 71, 97
Jekyll Island, 196n12
Jesuit, 17, 184n1, 194n2
Joe Bell site, 161, 166, 168
Johns Mound, 159, 163
Jordan River, 188n1
Jororo, 94, 97
Joseph de León site, 191n14

Juanillo revolt, 17, 26, 94, 95. *See also* revolts
Jumano, 51

Kasihta, 195n2. *See also* Cussita
Kent Mound, 28, 160, 165, 166
Killearn Borrow Pit, 28, 118, 161, 166, 168
kin, 75, 98, 125, 140, 196n10
kin groups, 57
King site, 161, 167, 168
kin selection, 57, 58
kinship, 47, 72, 141, 144, 154, 155, 174
kin structured migration, 23
Kiowa, 51
Koasatis, 146

La Chua Ranch, 95, 189n2
La Concepción de Ayubale, 192n14
La Florida Bioarchaeology Project, 4
Lake George, 97
Lake Jackson Mounds, 28, 111, 118, 123, 160, 165, 166
Lamar tradition, 113
La Salle, 95, 96, 189n2
La Tama (province), 87, 184n3, 193n19
Laudonniére, Rene, 72, 73, 74, 118, 185n6
Leake site, 161, 167, 168
LEH. *See* linear enamel hypoplasia
Le Moyne, Jacques, 72, 74, 118
Lewis Creek Mounds, 28, 160, 166
life course, 5, 109, 110, 111, 119, 120, 126, 144, 179
life cycle, 3, 10, 43, 44, 179
lifestyles, 2, 4, 16, 31, 109, 126
linear enamel hypoplasia, 17, 121, 123, 124, 125
linguistic boundary, 3, 114, 158
linguistic diversity, 40, 69
Little Egypt, 160, 164, 165, 166
Little Pine Island, 28, 160, 166
Lost Colony, 78. *See also* Roanoke
Lower Creek: Apalachee intermarriage with, 144; divergence from Seminole, 52, 130, 132, 135, 150, 195n6; early history, 132, 133, 135, 136, 137; English relations, 137, 152, 195n4; ethnic composition of, 141; later history, 134; movement back to Florida, 137, 139, 152, 153; political leanings, 140; treaty negotiations, 195n5; and the Yamassee, 142
Lucayos Indians, 188n1

maize agriculture, 13, 14, 68, 110, 157
maize consumption, 16, 125. *See also* diet
majolica, 25, 194n3
major genes, 19
Malica, 185n7
Manchester School, 63
María de la Cruz site, 111, 191n14
marine resources, 14, 16, 125
maroons, 51, 98, 136
Marxism, 3, 102
Marys Mound, 159, 163
Maskókî, 154, 155, 156, 172
matchlock, 90, 191n14, 192n17, 192n18, 193n18. *See also* firearms
mate exchange, 20, 40, 41, 45, 46, 70, 177, 183n1; long distance, 71, 165, 173; network, 9, 22, 23, 42, 94, 173; proscriptions against, 177; and social identity, 52, 56, 57, 58, 67, 174, 175, 182. *See also* gene flow
material culture, 3, 37; and the body, 53, 54, 55, 175; and identity, 44, 56, 57, 58, 63, 80, 173; mission inventories, 111, 114, 191n14
Mayaca, 71, 94, 97, 193n19
Mayport Mound, 159, 163
McGillivray, Alexander, 134, 135, 195n6
McKeithen Mound C, 159, 163
McLeod Mound, 159, 163
Menéndez de Avilés, Pedro, 13, 17, 71, 72, 73, 184n1, 188n5, 189n1
mesiodistal, 8, 9, 28, 29, 162, 165

mestizo, 60, 61, 111. *See also* mixed offspring
Miccosukees, 147, 155. *See also* Mikasuki
microdifferentiation, 20, 42
microevolution, 2, 5, 6, 8, 9, 12, 18, 19, 23, 34, 35, 37, 46, 52, 56, 75, 173, 175, 176, 181. *See also* gene flow; genetic drift; migration
microwear, 16
Middle Passage, 87. *See also* Africa; slaves; West Africa
middle-range theory, 2, 4, 174
migration, 18, 20, 22, 39, 53, 57, 93, 106, 131, 164, 170, 171, 180; as explanation of material culture patterns, 114; during the early 17th century, 177; during the later 17th century, 126, 179, 180; forced, 190n13; and fugitivism, 156, 157; and identity, 62; long distance, 41, 66, 193n19; microevolutionary effects of, 23, 35, 40, 41, 42, 175; rationale for, 10, 130, 132, 145, 150, 179; of the Seminole, 135, 136, 138, 139, 140, 143, 144, 146, 150, 151, 152, 179; and Timucua language origins, 185n9; transnational, 63. *See also* gene flow
Mikasuki, 136. *See also* Miccosukees
minority, 50, 112, 147, 148, 149
miserables, 82
missing data, 7, 8, 161, 163, 164, 166
missionization, 12, 13, 16, 17, 42, 96, 122, 123, 126, 157
mission list, 27, 194n5
Mississippian, 68, 69, 110, 130, 151, 157, 176
Mississippi River, 146, 189n2
mixed offspring, 54, 58, 114. *See also mestizo*
Mobile, 189n2
Mocama, 112, 141, 185n7, 193n19
Mococo, 185n7
model-bound, 7, 18, 19, 161
model-free, 18, 20

Molona, 185n7
monogamy, 72, 127
Moore, James, 17, 88, 96, 128, 129, 132, 140, 141, 143, 144, 179, 192n17
morbidity, 9, 16, 123, 125, 126, 176
mortality, xiii, 9, 26, 115, 116, 123, 127, 132, 156, 177
mortuary practices, 37, 114, 115, 118, 124, 126, 158, 179. *See also ad sancto*
Mucius Scaevola, 1, 183
multifactorial inheritance, 181
multilingual, 92, 154. *See also* polyglot
multivariate statistics, 18, 19, 20, 161, 171
Muscogulge, 147
Muskogee (language), 13, 14, 146, 184n5, 185n9
Muskogee (people), 132, 146, 147, 148, 149, 150, 163
mutation, 9, 175

Napa, 185n7
Napituca, 185n7
Napochie, 73
narrow sense heritability, 8, 19, 20, 22, 24, 29, 35, 175
Narváez, Pánfilo de, 13, 17, 118, 188n4, 188n7
national identity, 2, 47, 48, 49, 63, 187n6
nativism, 134, 135
natural selection, 9, 19, 22, 42, 57, 175
neo-Chumash, 51
nepotism, 57
New Orleans, 189n2
nitrogen isotopes, 16
Nombre de Dios, 115
nomocide, 129, 150, 151
nonrandom migration, 23
Norman Mound, 28, 160, 166
nursing, 109

observer error, 7, 28, 186n13
Ocale, 185n7
Ocaneechies, 88, 89
Ocheses, 91. *See also* Uchise

Index   **241**

Ochlockonee River, 13
Ocmulgee River, 96, 132
Ocone, 185n8
Oconee River, 159, 161, 166
Oconees, 140, 145, 147, 148, 150, 153, 193n19, 196n12. *See also* San Francisco de Oconee
O'Connell mission site, 118, 192n14
Ocute, 142
odontometrics, 6, 7, 8, 19, 28, 175. *See also* tooth size
Ogeechee River, 13, 28, 160, 184n3
Oglethorpe, James, 137, 196n8
Omittaguq, 185n7
Omoloa, 185n7
Onachaquara, 185n7
Onatheaqua, 74, 185n7
Orista, 75, 184n2, 184n3
Ortíz, Juan, 118, 188n4
Osceola, 136
Osochis, 147
Ossuary at Santa Catalina, 24, 25, 26, 27, 34, 122; effective population size of, 31; R-matrix analysis of, 31, 32
osteoarthritis, 18, 122, 125
osteological paradox, 125
Outina, 185n7
Oviedo, 183

Pacaha, 72, 74, 188n6
Pacaras, 148, 193n19
pandemic, 149, 168
panethnic, 10, 78, 79, 82, 92, 100
panethnogenesis, 50, 98
Pansacola, 95
Pardo, Juan, 194n21
pathology, 4, 17, 119, 120, 121, 123, 176. *See also* cribra orbitalia; porotic hyperostosis
Patica, 185n7
peaches, 127
Penn, William, 95, 190n8
peopling of Polynesia, 53
peopling of the New World, 53

periostitis, 18, 121, 123, 124
Pine Harbor, 122, 161, 162, 168
pirates, 95, 189n3, 193n18
plantation, 51, 85, 98, 189n4, 190n5, 190n10
plural wives, 109
Pocotaligo, 133
Pohoy, 97, 185n7
polar teeth, 8, 9, 183n4
polyethnic, 51, 97, 133, 144, 179
polygamy. *See* plural wives
polygenic, 7, 181
polyglot, 14, 155. *See also* multilingual
polygyny. *See* plural wives
population aggregation: in Apalachee province, 148; as explanation of material culture patterns, 114; during the formation of the Creeks, 146, 149; during the formation of the Seminole, 153; as a response to epidemics, 123; Spanish policies of, 62, 80, 82, 83, 110, 178
population boundary, 45, 64
population, definition of, 40, 45
population history, 5
population origins, 53, 57
population replacement, 35
population size, 9, 18, 23; in Apalachee province, 194n4; census, 185n11, 186n11; decline in, 14, 16, 22, 23, 31, 39, 40, 41, 42, 82, 94, 123, 156, 176, 183n1; effective, 19, 20, 22, 29, 41, 175, 185n11, 186n11; estimates of, 31, 32, 41, 123, 148, 156. *See also* demographic collapse
population structure, 5, 9, 18, 22, 23, 28, 150, 168; during the early 17th century, 12, 29, 30, 41, 45; during the later 17th century, 32, 33, 42, 78, 126, 150, 177; during the precontact period, 29, 163, 164, 166, 167; and effective population size, 186n11. *See also* isolation-by-distance
porotic hyperostosis, 17, 121, 124, 125
Port Royal chiefdoms, 184n3

posture, 179
Potano, 95, 113, 185n7, 185n8. *See also* San Francisco de Potano
pottery. *See* ceramics; ceramic style
Powhatan, 75, 78, 85, 86, 87, 95, 96
practice theory, 10, 49, 51, 54, 80, 100, 101, 176, 194n1; and ethnic identity, 103, 104, 105, 107, 108, 119, 120, 127
preservation (skeletal), 2, 7, 26, 125, 161
Prieto, Father Martín, 188n7
primodial charter, 153
primordialism, 49, 57, 58, 80, 101, 102, 103, 187n5
proto-Creek, 148, 178, 180
proto-Seminole, 10, 130, 133, 135, 137, 138, 139, 140, 141, 143, 144, 145, 147, 148, 150, 151, 152, 153, 157, 170, 179, 180, 195n5, 195n6

quantitative genetics, 6, 10, 22, 29, 161
Queen Anne's War, 88

race, 7, 49, 56, 57, 58, 77, 82, 173, 174
race science, 5, 183n2, 186n3
raiding, 16, 46, 84, 172, 179, 190n11, 190n12; between Apalachee and Timucua, 188n7; by the Cherokee, 137; during the early 17th century, 82, 83, 97, 98, 194n21; by James Moore, 128, 129, 132, 140, 141, 143; by pirates, 95; as a precursor to Seminole migration, 145, 152, 153; among protohistoric chiefdoms, 74.
Ranjel, Rodrigo, 1, 72, 183, 188n6, 188n7
rebellions. *See* revolts
Rebolledo, Governor, 126, 192n16
redistribution: of goods, 68; of people, 69, 157, 176
Red Stick Creeks, 134, 135
refugees, 91, 112, 141, 146, 147, 148, 149
reification, 4, 43, 45, 62, 63, 64, 80, 83, 100, 129, 153, 174, 178, 183n2, 187n4
religious identity, 47, 115, 173

repartimiento, 16, 18, 82, 84, 86, 91, 112, 125, 127, 178, 194n6
república de españoles, 81
república de indios, 81
Republic of Indians, 82, 83, 84, 126, 179
revolts, 16, 17, 66, 82, 88, 93, 94, 95, 96, 190n10, 194n6. *See also* Juanillo revolt; Timucua, rebellion of 1656
rice, 86
Richardson site, 113
R-matrix 19, 20, 22, 29, 30, 31, 32, 33, 40, 41, 175
RMET, 19, 29, 30, 32, 33
Roanoke, 85. *See also* Lost Colony

Sabacola, 142, 143, 190n12
sacred fire, 143
Salazar, Governor Hita, 192n16
Saltwater district, 112
sample bias, 19, 34, 35, 120, 125, 161
sample size, 6, 7, 27, 29, 161, 162, 164, 166
San Agustín de Urica, 113, 191n14
San Antonio Anacape, 193n19
San Carlos de Yatcatani, 83
San Damián de Escambé, 115, 191n14
San Francisco de Oconee, 192n14, 196n12
San Francisco de Potano, 191n14
San Joseph de Sapala, 87, 95
San Juan de Aspalaga, 192n14
San Juan de Guacara, 88, 96, 113
San Juan del Puerto, 26, 112, 191n14
San Lorenzo de Ivitachuco, 95, 115
San Luis de Talimali, 1, 24, 25, 91, 179; burial practices, 115, 118, 125; effective population size of, 32; firearms at, 191n14; health, 122, 125, 194n5; R-matrix analysis of, 32, 33, 34, 35
San Martín de Timucua, 24, 25, 26, 27, 30, 31, 194n3; burial practices, 115, 118; ceramics at, 112, 113; effective population size of, 31; firearms at, 191n14; R-matrix analysis of, 31, 32, 34, 35
San Miguel de Asile, 115, 186n12

Index 243

San Miguel de Gualdape, 17
San Nicolás de Tolentino, 82
San Pedro y San Pablo de Patale, 24, 25, 122; burial practices, 115, 116, 123, 124; effective population size of, 31; firearms at, 192n14; health, 122, 124; R-matrix analysis of, 30, 31, 32, 34, 35
San Pedro y San Pablo de Potohiriba, 26, 115
San Salvador de Mayaca, 193n19
Santa Catalina de Ahoica, 88, 95
Santa Catalina de Guale, 24, 25, 26, 27; burial practices, 115, 116, 124, 125; effective population size of, 31; health, 122, 123, 124, 125; R-matrix analysis of, 31, 32, 33, 34, 35
Santa Catalina de Guale de Santa María, 24, 25, 26, 27; burial practices, 115, 116, 118; effective population size of, 32; health, 122, 125; R-matrix analysis of, 31, 33, 34, 35
Santa Elena, 189n1, 189n5
Santa Fé de Toloca, 26, 88, 96, 115
Santa María de los Yamassee, 24, 25, 27, 122; effective population size of, 32; R-matrix analysis of, 33, 34, 35
Santa María Island, 26, 33
Santiago (translator), 185n10
Santo Domingo de Talaxe, 87, 95
Sapelo River, 28, 159
Sattler, Richard, xiii, xiv, 131, 139, 140, 141, 142, 145, 155, 196n9
Saturiwa, 185n7
Savannah (people), 89, 133
Savannah River, 28, 87, 95, 160, 184n3, 195n4
Sawoklis, 147
Scott Miller site, 192n14
Searle, Robert, 17, 95, 193n18
Seaside Mounds, 159, 163
Secoffee, 138, 140, 142, 196n10
Second Seminole War, 135, 195n7
sedentary, 13, 68, 82, 110
sedentism, 111

selection. *See* kin selection; natural selection
Seloy, 185n7
Seminole: as descendants of precontact Floridians, 11, 157; divergence from Lower Creeks, 52, 134, 136, 145, 150, 171, 172, 180, 195n6, 195n7; ethnogenesis of, xiii, 10, 130, 131, 143, 144, 148, 151, 153, 169, 170, 172, 179, 180, 195n7; etymology, 149; as an identity, 2, 127, 138, 148, 150, 195n4, 195n7; migration rationale, 137, 138, 140, 146, 147, 152, 155; migrations, 130; Oklawaha band, 142; oral tradition, 139, 155; period of Florida history, xiv, 72, 151; traditional historical views on, 128, 129, 132, 134, 135, 136, 154, 179; village composition, 135. *See also* Alachua Seminole; Apalachee Seminole; proto-Seminole
Senquene, 188n1
sepulturies, 116
settlement Indians, 93
settlement structure, 14, 37, 42, 68, 110
Seven Mile Bend, 28, 160, 165, 166
shatter zone, 93. *See also* tribal zone
Shawnee, 88, 133, 135, 146
Siminioles, 149
Simpukasi, 140. *See also* Secoffee
Sioux (people), 51
Siouxan (language), 185n9
sistema de castas, 54, 126
Sixtoe Field, 159, 163
slave labor, 2, 16, 26, 54, 86, 77, 125, 127, 178, 185n7, 192n17, 195n3; as a cause of migration, 148; during the early 16th century, 189n5, 190n5; and identity, 102, 107, 129; targeting Christians, 91; and the tribal zone, 93, 96; in West Africa, 190n9; Westo, 87, 88, 89, 90, 112, 192n15, 192n16
slaves, 190n10; African importation of, 87, 137, 189n4, 190n10; among Native American groups, 140; exporting of,

86; Native American, 86, 87, 97, 99, 137, 143, 179, 190n7, 190n10
smallpox, 149, 195n3. *See also* epidemics
Snow Beach, 28, 118, 161, 166, 168
social constructionism, 56
sociobiology, 56, 57, 58, 174
sociopolitical boundary, 3, 66, 76, 92, 158, 162
solidarity, 48, 54, 58, 59, 103, 104, 106, 108, 109, 114, 115, 126, 174, 175, 178, 179, 187n4
South Carolina, 70, 73, 86, 87, 96, 193n19
South End Mound, 28, 160, 166
standard error, 19; analytical results, 30, 31, 32, 33
state expansion, 94, 108
statistical scaling, 20, 22, 29, 30, 31, 33, 34, 35, 41, 162
status, 1, 48, 61, 63, 64, 71, 72, 84, 143; identity, 47, 80, 107, 173
St. Augustine, 16, 26, 62, 84, 129, 133, 175, 179; African slaves in, 189n4; aggregation at, 61, 82, 89, 91, 126; and Seminole ethnogenesis, 137, 140; assaults on, 17, 88, 95, 96, 189n3, 193n18; ceramic inventory, 111, 112; economic importance of, 68; firearms in, 90, 190n14, 191n14, 192n17, 193n18; founding of, 13, 17, 189n2; revolts, 95; under English control, 195n4
St. Catherines Island, 26, 28, 88, 90, 95, 159, 160, 161, 189n1, 192n15
St. Johns River, 13, 82, 159, 163, 165, 170, 175, 185n6, 189n1
St. Marks River, 137, 189n2
St. Simons Island, 28, 88, 95, 159, 160
stinkard, 147
striae of Retzius, 17
Stuart's Town, 84, 95, 96
style (material culture), 3, 44, 54, 55, 56, 58, 111, 114, 176
subsistence, 14, 37, 110, 138, 145. *See also* diet

sugarcane, 190n8
suicide, 90
superordinate identity, 81, 82, 83

Tacatacuru, 185n7
Tallahassee, 1, 138, 139, 140, 195n4
Tama (people), 148
Tamathlis, 147
Tampa Bay, 70, 165
Tasquique, 97, 142, 143
Tasquiqui (town), 195n2
Tatham Mound, 28, 161, 166, 168
taxation. *See repartimiento*
Tecumseh, 134, 135
Tennessee Valley, 194n21
Tequesta, 72, 73
terminus ante quem, 25, 26
Tierra Verde Mound, 28, 160, 165, 166
Timucua, 7, 11, 145, 148, 194n3; analytical results for, 42, 163, 164, 165, 166, 170; attacks directed against, 17, 83, 88, 91, 96, 97, 99; burial practices, 118; ceramics, 111, 112, 113; eastern province, 75; emigration to, 82; establishing missions among, 13; ethnographic imagery of, 118; etymology, 184n6; firearms, 191n14; French occupation, 72, 73, 74; fugitivism, 193n19; health, 122; as an identity, 61, 65, 66, 81, 101, 108, 114, 128, 130, 144, 150, 172; kinship among chiefs, 72; language, 14, 185n8, 185n9, 185n10; political structure, 75, 185n7; population size, 31, 194n5; post-1700, 141; as proto-Seminole, 130, 131, 143, 147, 155, 156; rebellion of 1656, 17, 26, 84, 95, 96, 113, 126, 191n14, 192n16, 193n19; refugees in, 91, 193n19; revolts, 95; settlement organization, 111; skeletal samples, 24, 25, 26, 27, 148, 159, 160, 163, 166, 177; subsistence, 14; warfare, 17, 74, 188n7; western province, 70. *See also* San Martín de Timucua
tobacco, 86, 190n8

Tocobaga, 27, 73, 97, 148, 160, 161, 185n7
Tomahitans, 88
Tonapi, 140, 195n4, 196n9. *See also* Tunapé
tooth size, 6, 29, 162, 164, 165, 166, 167, 169, 175; attrition effects on, 28; benefits of using, 7; and body size, 9; environmental effects on, 8; narrow sense heritability of, 8; and studies of identity, 55, 173. *See also* odontometrics
trace elements analysis, 16
trade: Apalachee-Havana, 189n2; Creek and English, 132; as an explanation for ceramic distribution, 112, 114; long distance, 68; as a motivator for English expansion, 178; among precontact chiefdoms, 69, 154. *See also* fur trade; firearms, illegal trade in
trade beads, 118
trade goods, 66, 68, 69
traders, 16, 51, 71, 84, 87, 90, 195n6
translators. *See* Santiago
Treaty of Moultrie Creek, 134, 195n7
Treaty of New York, 134
Treaty of Paynes Landing, 134
Treaty of Pensacola, 195n5
Treaty of Picolata, 195n5, 195n7
tribal boundary, 66
tribal identity, 10, 62, 136, 162, 172, 179
tribalization, 10, 51, 64, 65, 66, 68, 79, 85, 93, 94, 96
tribal zone, 4, 10, 64, 77, 78; and conflict, 94, 96, 102; definition of, 79, 193n20; during the early 17th century, 65, 66, 68; during the later 17th century, 78, 81, 91, 92, 93, 101, 178
tribe-state interaction, 79, 93
Trinity Episcopal Church site, 191n14
Tuckabatchees, 146, 147
Tucuru, 185n7
Tucururu, 185n7

Tulafina, 184n3
Tunapé, 140. *See also* Tonapi
Tuscaroras, 78, 88
typology, 5, 7, 53, 57, 174, 184n2

Uçachile, 188n7
Uchise, 16, 142, 146, 148, 172, 195n4; raid on missions, 87, 95, 96
Ucito, 185n7
uniformitarian, 3
Upper Creek, 134, 135
Uriutina, 185n7
Utina, 26, 95, 113, 159, 185n7, 194n3
Venables, Robert, 95, 190n8
Verrazano, Giovanni da, 190n5

Waddells Mill Pond, 28, 161, 166, 168
Warao, 185n9
warfare, 3, 51, 64, 65, 94, 97, 154, 192n18; Apalachee-Timucua, 188n7; biological effects of, 73, 74, 75, 177; internecine, 66, 76, 77, 96; in Mississippian chiefdoms, 69, 73, 76, 98, 176;
wear. *See* attrition, dental
Weeki Wachee, 28, 161, 166, 168
West Africa, 178, 190n9
Westo, 17, 84, 89, 90, 91, 96, 98, 99; decline of, 88; raid on Guale, 83, 87, 95, 112, 192n15, 193n19
Wickman, Patricia, 129, 131, 154, 155, 156, 157, 172
Woodlief site, 161, 166, 168
World Systems Theory, 4, 178, 190n9
worldview, 73, 131, 155, 156. *See also* cosmogony

Yamacraws, 147
Yamassee, 27, 87, 143; as Christians, 130, 133, 144, 177, 193n19; in La Florida, 84, 142, 156, 172; raids on missions, 88, 89, 91, 95, 96, 99, 190n11; and Seminole ethnogenesis, 140, 141, 142, 143, 144, 145, 147, 148, 150, 155;

skeletal samples, 177; as subjects of raids, 96. *See also* Santa María de los Yamassee
Yamassee War, 84, 133, 137, 142, 195n3

Ybarra, Governor, 185n10
Yuchi, 133, 146, 155

Yufera, 185n7, 185n8
Yui, 185n7
Yustaga, 75, 84, 113, 185n7, 186n12, 188n7, 193n19

Zetrouer site, 113
z-score transformation, 37

Christopher M. Stojanowski is a bioarchaeologist affiliated with the Center for Bioarchaeological Research at Arizona State University's School of Human Evolution and Social Change. He is the author of *Biocultural Histories in La Florida: A Bioarchaeological Perspective* and is the coeditor of *Bioarchaeology and Identity in the Americas*.

Ripley P. Bullen Series
Florida Museum of Natural History

Tacachale: Essays on the Indians of Florida and Southeastern Georgia during the Historic Period, edited by Jerald T. Milanich and Samuel Proctor (1978)

Aboriginal Subsistence Technology on the Southeastern Coastal Plain during the Late Prehistoric Period, by Lewis H. Larson (1980)

Cemochechobee: Archaeology of a Mississippian Ceremonial Center on the Chattahoochee River, by Frank T. Schnell, Vernon J. Knight Jr., and Gail S. Schnell (1981)

Fort Center: An Archaeological Site in the Lake Okeechobee Basin, by William H. Sears, with contributions by Elsie O'R. Sears and Karl T. Steinen (1982)

Perspectives on Gulf Coast Prehistory, edited by Dave D. Davis (1984)

Archaeology of Aboriginal Culture Change in the Interior Southeast: Depopulation during the Early Historic Period, by Marvin T. Smith (1987)

Apalachee: The Land between the Rivers, by John H. Hann (1988)

Key Marco's Buried Treasure: Archaeology and Adventure in the Nineteenth Century, by Marion Spjut Gilliland (1989)

First Encounters: Spanish Explorations in the Caribbean and the United States, 1492–1570, edited by Jerald T. Milanich and Susan Milbrath (1989)

Missions to the Calusa, edited and translated by John H. Hann, with an introduction by William H. Marquardt (1991)

Excavations on the Franciscan Frontier: Archaeology at the Fig Springs Mission, by Brent Richards Weisman (1992)

The People Who Discovered Columbus: The Prehistory of the Bahamas, by William F. Keegan (1992)

Hernando de Soto and the Indians of Florida, by Jerald T. Milanich and Charles Hudson (1993)

Foraging and Farming in the Eastern Woodlands, edited by C. Margaret Scarry (1993)
Puerto Real: The Archaeology of a Sixteenth-Century Spanish Town in Hispaniola, edited by Kathleen Deagan (1995)

Political Structure and Change in the Prehistoric Southeastern United States, edited by John F. Scarry (1996)

Bioarchaeology of Native Americans in the Spanish Borderlands, edited by Brenda J. Baker and Lisa Kealhofer (1996)

A History of the Timucua Indians and Missions, by John H. Hann (1996)

Archaeology of the Mid-Holocene Southeast, edited by Kenneth E. Sassaman and David G. Anderson (1996)

The Indigenous People of the Caribbean, edited by Samuel M. Wilson (1997; first paperback edition, 1999)

Hernando de Soto among the Apalachee: The Archaeology of the First Winter Encampment, by Charles R. Ewen and John H. Hann (1998)

The Timucuan Chiefdoms of Spanish Florida, by John E. Worth: vol. 1, *Assimilation*; vol. 2, *Resistance and Destruction* (1998)

Ancient Earthen Enclosures of the Eastern Woodlands, edited by Robert C. Mainfort Jr. and Lynne P. Sullivan (1998)

An Environmental History of Northeast Florida, by James J. Miller (1998)

Precolumbian Architecture in Eastern North America, by William N. Morgan (1999)

Archaeology of Colonial Pensacola, edited by Judith A. Bense (1999)

Grit-Tempered: Early Women Archaeologists in the Southeastern United States, edited by Nancy Marie White, Lynne P. Sullivan, and Rochelle A. Marrinan (1999)

Coosa: The Rise and Fall of a Southeastern Mississippian Chiefdom, by Marvin T. Smith (2000)

Religion, Power, and Politics in Colonial St. Augustine, by Robert L. Kapitzke (2001)

Bioarchaeology of Spanish Florida: The Impact of Colonialism, edited by Clark Spencer Larsen (2001)

Archaeological Studies of Gender in the Southeastern United States, edited by Jane M. Eastman and Christopher B. Rodning (2001)

The Archaeology of Traditions: Agency and History Before and After Columbus, edited by Timothy R. Pauketat (2001)

Foraging, Farming, and Coastal Biocultural Adaptation in Late Prehistoric North Carolina, by Dale L. Hutchinson (2002)

Windover: Multidisciplinary Investigations of an Early Archaic Florida Cemetery, edited by Glen H. Doran (2002)

Archaeology of the Everglades, by John W. Griffin (2002)

Pioneer in Space and Time: John Mann Goggin and the Development of Florida Archaeology, by Brent Richards Weisman (2002)

Indians of Central and South Florida, 1513–1763, by John H. Hann (2003)

Presidio Santa Maria de Galve: A Struggle for Survival in Colonial Spanish Pensacola, edited by Judith A. Bense (2003)

Bioarchaeology of the Florida Gulf Coast: Adaptation, Conflict, and Change, by Dale L. Hutchinson (2004)

The Myth of Syphilis: The Natural History of Treponematosis in North America, edited by Mary Lucas Powell and Della Collins Cook (2005)

The Florida Journals of Frank Hamilton Cushing, edited by Phyllis E. Kolianos and Brent R. Weisman (2005)

The Lost Florida Manuscript of Frank Hamilton Cushing, edited by Phyllis E. Kolianos and Brent R. Weisman (2005)

The Native American World Beyond Apalachee: West Florida and the Chattahoochee Valley, by John H. Hann (2006)

Tatham Mound and the Bioarchaeology of European Contact: Disease and Depopulation in Central Gulf Coast Florida, by Dale L. Hutchinson (2006)

Taino Indian Myth and Practice: The Arrival of the Stranger King, by William F. Keegan (2007)

An Archaeology of Black Markets: Local Ceramics and Economies in Eighteenth-Century Jamaica, by Mark W. Hauser (2008)

Mississippian Mortuary Practices: Beyond Hierarchy and the Representationist Perspective, edited by Lynne P. Sullivan and Robert C. Mainfort Jr. (2010; first paperback edition, 2012)

Bioarchaeology of Ethnogenesis in the Colonial Southeast, by Christopher M. Stojanowski (2010; first paperback edition, 2013)

French Colonial Archaeology in the Southeast and Caribbean, edited by Kenneth G. Kelly and Meredith D. Hardy (2011)

Late Prehistoric Florida: Archaeology at the Edge of the Mississippian World, edited by Keith Ashley and Nancy Marie White (2012)

Early and Middle Woodland Landscapes of the Southeast, edited by Alice P. Wright and Edward R. Henry (2013)

www.ingramcontent.com/pod-product-compliance
Lightning Source LLC
Chambersburg PA
CBHW031433160426
43195CB00010BB/712